ASYMPTOTIC THEORY OF
STATISTICAL TESTS AND ESTIMATION

Wassily Hoeffding

ASYMPTOTIC THEORY OF STATISTICAL TESTS AND ESTIMATION

IN HONOR OF WASSILY HOEFFDING

Edited by

I. M. Chakravarti

Department of Statistics
University of North Carolina at
Chapel Hill, North Carolina

ACADEMIC PRESS

A Subsidiary of Harcourt Brace Jovanovich, Publishers

New York London. Toronto Sydney San Francisco 1980

Academic Press Rapid Manuscript Reproduction

The proceedings of an advanced international symposium held at the University of North Carolina at Chapel Hill on April 16–18, 1979.

ACADEMIC PRESS, INC.
111 Fifth Avenue, New York, New York 10003

United Kingdom Edition published by
ACADEMIC PRESS, INC. (LONDON) LTD.
24/28 Oval Road, London NW1 7DX

Library of Congress Cataloging in Publication Data

Advanced International Symposium on Asymptotic Theory of Statistical Tests and Estimation, University of North Carolina at Chapel Hill, 1979.
Asymptotic theory of statistical tests and estimation.

"Sponsored by the Department of Statistics of the University of North Carolina at Chapel Hill and the Institut de statistique des universités de Paris."
"Published works of Wassily Hoeffding: p.
1. Statistical hypothesis testing—Asymptotic theory—Congresses. 2. Estimation theory—Asymptotic theory—Congresses. 3. Höffding, Wassilij. I. Höffding, Wassilij. II. Chakravarti, Indra Mohan. III. University of North Carolina at Chapel Hill, Department of Statistics. IV. Institut de statistique des universités de Paris. V. Title.
QA277.A38 1979 519.5′6 80-10267
ISBN 0-12-166650-6

PRINTED IN THE UNITED STATES OF AMERICA

80 81 82 83 9 8 7 6 5 4 3 2 1

Contents

16 April 1979

*Speakers at the symposium.

17 April 1979

18 April 1979

Contributors

Numbers in parentheses indicate the pages on which the authors' contributions begin.

R. R. Bahadur *(33), Department of Statistics, University of Chicago, Chicago, Illinois 60637*

C. B. Bell *(127), Biostatistics SC–32, University of Washington, Seattle, Washington 98195*

P. J. Bickel *(307), Department of Statistics, University of California, Berkeley, California 94720*

N. Bönner *(85), Institut für Mathematische Stochastik des Albert–Ludwigs Universität, 78 Freiburg im Breisgau, Hermann Herder Strasse 10, D 7800, West Germany*

Jacques Chevalier *(231), Institut de Statistique des Universités de Paris, 4 Place Jussieu–T. 45-55–E. 3, 75230 Paris 05, France*

Paul Deheuvels *(259), Institut de Statistique des Universités de Paris, 4 Place Jussieu–T. 45-55–E. 3, 75230 Paris 05, France*

Daniel Dugué *(65), Institut de Statistique des Universités de Paris, 4 Place Jussieu–T. 45-55–E. 2, 75230 Paris 05, France*

Jean Geffroy *(159), Institut de Statistique des Universités de Paris, 4 Place Jussieu–T. 45-55–E.3, 75230 Paris 05, France*

J. C. Gupta *(33), Indian Statistical Institute, 203 Barrackpore Trunk Road, Calcutta 35, India*

W. J. Hall *(325), Department of Statistics, University of Rochester, Rochester, New York 14627*

Eugene Lukacs *(205), Catholic University, 3727 Van Ness Street NW, Washington, DC 20016*

U. Müller-Funk *(85), Institut für Mathematische Stochastik des Albert–Ludwigs Universität, 78 Freiburg im Breisgau, Hermann Herder Strasse 10, D 7800, West Germany*

Jerzy Neyman *(1), Department of Statistics, University of California, Berkeley, California 94720*

F. Ramirez *(127), Biostatistics SC–32, University of Washington, Seattle, Washington 98195*

James Reeds *(287), Department of Statistics, University of California, Berkeley, California 94720*

P. Révész *(147), Mathematical Institute of the Hungarian Academy of Sciences, Reáltanoda University 13–15, Budapest, H–1053, Hungary*

Herbert Robbins *(251), Department of Applied Mathematics and Statistics, State University of New York, Stonybrook, New York 11794*

B. W. Silverman *(179), School of Mathematics, University of Bath, Claverton Down, Bath BA2 7AY, England*

Eric Smith *(127), Biostatistics SC–32, University of Washington, Seattle, Washington 98195*

W. R. van Zwet *(307), Department of Mathematics, University of Leiden, Wassenaarsweg 80, Postbus 9512, 2300 RA Leiden, The Netherlands*

H. Witting *(85), Institut für Mathematische Stochastik des Albert–Ludwigs Universität, 78 Freiburg im Breisgau, Hermann Herder Strasse 10, D 7800, West Germany*

S. L. Zabell *(33), Department of Statistics, University of Chicago, Chicago, Illinois 60637*

Preface

This volume contains the proceedings of the Advanced International Symposium on Asymptotic Theory of Statistical Tests and Estimation held at the University of North Carolina at Chapel Hill, April 16–18, 1979, in honor of Wassily Hoeffding. The symposium was sponsored by the Department of Statistics of the University of North Carolina at Chapel Hill, and the Institut de Statistique des Universités de Paris, with support from the National Science Foundation under Grant No. MCS-78 24 665 and from the Department of the Navy through the Office of Naval Research under Grant No. N00014-79-G-0020. We thank these agencies, and in particular Dr. E. J. Wegman of the Office of Naval Research and Drs. W. H. Pell and B. E. Trumbo of the National Science Foundation, for their invaluable support.

In consultation with colleagues, the editor, who acted also as the principal investigator, invited prominent research workers in the areas of asymptotic theory of statistical tests and estimation from the United States and Europe to speak at the symposium and to contribute papers, based on their talks, to this volume.

Dean S. R. Williams of the College of Arts and Sciences kindly gave the welcoming address and Professor J. Neyman the opening address on April 16. The first day of the symposium happened to be Professor Neyman's eighty-fifth birthday, and there was a birthday celebration between sessions.

On the initiative of Professor Neyman the following message was sent to Professor Harald Cramér on behalf of all present at the symposium: "The invited speakers at the Advanced International Symposium on Asymptotic Theory of Statistical Tests and Estimation now going on at Chapel Hill in honor of Wassily Hoeffding have made repeated reference to your pioneering work on large deviations, which has opened up so many new directions in statistical theory. They and the audience at large want to take this opportunity to convey to you their gratitude for your many past contributions and to express their hopes for the continuance *ad infinitum* of your presence and influence among them."

Papers presented at the symposium treated large deviations theory, distributions—exact and asymptotic, extreme multivariate distributions, sequential analysis, probability inequalities, estimation of density, support and contours of support of probability laws, statistical inference for the compound Poisson and Wiener–Lévy processes, properties of Wiener processes, and applications.

Neyman traces the origin of the theory of large deviations in the work of Cramér, followed by that of Hoeffding. Bahadur, Gupta, and Zabell derive a crucial lower bound to the limit inferior of a probability sequence by an ingenious application of the Neyman–Pearson lemma, and point out how various lower bounds in the large deviations literature can be derived from the present one. Reeds obtains a multinomial large deviations correction term taking into account the differential topology of the Kullback–Liebler information number and the set in question.

Dugué uses characteristic functions to solve multivariate distribution problems, in particular, the distribution of the multivariate analogue of the von Mises–Smirnov statistic. Lukacs proves stability theorems in the characterizations for normal and degenerate distributions. Deheuvels treats distributions of infinite order, infinite divisibility, and representations of extreme multivariate distributions using dependence functions. Révész proves a result about the smallness of a process related to the Wiener process, and Geffroy introduces the concept of "decantation" in the context of asymptotic separation of two families of probability sequences and discusses its application to consistency in estimation. While Robbins treats estimation problems for compound Poisson distributions, Bell, Ramirez, and Smith obtain statistical inference procedures for Wiener–Lévy processes.

Silverman attempts a reconciliation between theory and practice in density estimation, with illustrations from engineering data. Chevalier gives results on estimation of support and contour-line of support of a probability law from a sample. Witting, Bönner, and Müller-Funk derive the order of the correction term of the Chernoff–Savage representation of the correlation statistic and discuss its application to sequential testing of independence. Hall introduces sequential minimum probability ratio tests which include SPRTs and permit control of weighted averages of the error probabilities.

Van Zwet, in a joint paper with Bickel, presents a generalization of Hoeffding's theorem on the dispersion of the sum of the number of successes from Poisson binomial trials to Poisson multinomial trials.

We thank the speakers, chairmen of sessions, and all others who contributed to the success of this symposium.

Biographical Sketch

Wassily Hoeffding

The Advanced International Symposium on Asymptotic Theory of Statistical Tests and Estimation was held April 16–18, 1979 in Chapel Hill to honor Wassily Hoeffding on the occasion of his sixty-fifth birthday.

Wassily Hoeffding was born on June 12, 1914 in Mustamäki, Finland (Gorkovskoye, USSR since 1940), although his birth certificate shows his place of birth as St. Petersburg (now Leningrad). His father (of Danish ancestry) was an economist, and his mother (née Vvedenskiĭ) had studied medicine. Both his grandfathers were engineers. His father's uncle, Harald Hoeffding, was a well known philosopher.

In 1918, the Hoeffding family left their home in Tsarskoye Selo (now Pushkin, USSR) for the Ukraine and, after traveling through scenes of civil war, finally left Russia in 1920 for Denmark.

Wassily Hoeffding first went to school in Denmark, 1920–1922, then in Hamburg until 1924, when the family moved to Berlin. Wassily and his two brothers attended an Oberrealschule—a high school which emphasized natural sciences and modern languages. He attended the Handelshochschule in Berlin during 1933–1934, and in 1934 he entered Berlin University to study mathematics. He took courses with Erhard Schmidt and Ludwig Bieberbach, a course on number theory with Alfred Brauer, who later joined the Department of Mathematics of the University of North Carolina at Chapel Hill, and a course on probability and statistics with Alfred Klose, a disciple of Richard von Mises.

In 1940, he was awarded a Ph.D. degree in Mathematics by the University of Berlin for a dissertation on the properties of bivariate probability distributions that are invariant under monotone transformations of the margins.

He served on the editorial staff of the *Jahrbuch über die Fortschritte der Mathematik,* Berlin, (1940–1945), and was a member of the Berliner Hochschulinstitut für Versicherungswissenschaft, (1941–1945).

In September 1946, he came to the United States. During the next winter, he attended courses given by A. Wald, J. Wolfowitz, and J. Neyman at Columbia University. In 1947 he was appointed Research Associate at the new Department of Mathematical Statistics of the University of North Carolina at Chapel Hill. He was appointed Lecturer in 1948, Assistant Professor in 1949, and Associate Professor in 1952. (He became a citizen of the United States in 1952.) He was appointed Professor in 1956, and a Kenan Professor in 1973.

The Institute of Mathematical Statistics named him Wald Lecturer in 1967 and President in 1969. In 1976, he was elected to the National Academy of Sciences. He is also a member of the International Institute of Statistics, and a Fellow of the Institute of Mathematical Statistics, and of the American Statistical Association. He has served twice (1958–1961, 1964–1967) as an Associate Editor of the *Annals of Mathematical Statistics*.

Wassily Hoeffding's principal work deals with probability limit theorems, large-sample behavior of statistical tests, U-statistics, probability inequalities, sequential analysis, approximation errors, and completeness of families of distributions. Hoeffding has made major contributions in these and other areas, and each one of his papers, presented in his superbly succinct style, has inspired further research. He has supervised over sixteen Ph. D. dissertations. Many of his former students are today regarded as leaders in their fields of research. Although now retired, he is still actively pursuing his research program. We hope and expect that he will continue to make important contributions for many more years.

This volume is dedicated to Wassily Hoeffding with respect, admiration, and affection.

I. M. Chakravarti

Chapel Hill
21 November 1979

Published Works

of Wassily Hoeffding

Masstabinvariante korrelationstheorie. *Schriften des Mathematischen Instituts und des Instituts für Angewandte Mathematik der Universität Berlin 5,* 179–233 (1940).

Masstabinvariante korrelationmasse für diskontinuierliche verteilungen. *Archiv für Mathematische Wirtschafts–und Sozialforschung 7,* 49–70 (1941).

Stochastische abhängigkeit und funktionaler zusammenhang. *Skandinavisk Aktuarietidskrift 25,* 200–227 (1942).

On the distribution of the rank correlation coefficient τ when the variates are not independent. *Biometrika 34,* 183–196 (1947).

A class of statistics with asymptotically normal distribution. *Annals of Mathematical Statistics 19,* 293–325 (1948).

(with Herbert Robbins) The central limit theorem for dependent random variables. *Duke Mathematical Journal 15,* 773–780 (1948).

A Nonparametric test of independence. *Annals of Mathematical Statistics 19,* 546–557 (1948).

'Optimum' nonparametric tests. "Proceedings of the Second Berkeley Symposium on Mathematical Statistics and Probability," pp. 83–92. University of California Press, Berkeley, California, 1951.

A combinatorial central limit theorem. *Annals of Mathematical Statistics 22,* 558–566, (1951).

The large-sample power of tests based on permutations of observations. *Annals of Mathematical Statistics, 23,* 169–192, (1952).

On the distribution of the expected values of the order statistics. *Annals of Mathematical Statistics 24,* 93–100, (1953).

A lower bound for the average sample number of a sequential test. *Annals of Mathematical Statistics 24,* 127–130, (1953).

(With S. S. Shrikhande) Bounds for the distribution of a sum of independent, identically distributed random variables. *Annals of Mathematical Statistics 26,* 439–449 (1955).

(With Joan Raup Rosenblatt) The efficiency of tests. *Annals of Mathematical Statistics 26,* 53–63 (1955).

The extrema of the expected value of a function of independent random variables. *Annals of Mathematical Statistics 26,* 268–275 (1955).

On the distribution of the number of successes in independent trials. *Annals of Mathematical Statistics 27,* 713–721 (1956).

The role of assumptions in statistical decisions. *Proceedings of the Third Berkeley Symposium on Mathematical Statistics and Probability 1,* 105–114 (1956).

(With J. Wolfowitz) Distinguishability of sets of distributions. *Annals of Mathematical Statistics 29,* 700–718 (1958).

Lower bounds for the expected sample size and the average risk of a sequential procedure. *Annals of Mathematical Statistics 31,* 356–368 (1960).

An upper bound for the variance of Kendall's 'tau' and of related statistics "Contributions to Probability and Statistics, Essays in Honor of Harold Hotelling," pp. 258–260. Stanford University Press, 1960.

On sequences of sums of independent random vectors, "Proceedings of the Fourth Berkeley Symposium on Mathematical Statistics and Probability," Vol. II, pp. 213–226. University of California Press, Berkeley, California, (1961).

Probability inequalities for sums of bounded random variables. *Journal of the American Statistical Association 58,* 13–31 (1963).

Lower bounds for the expected sample size of a sequential test. "Information and Decision Processes" (R. E. Machol, ed.), pp. 53–61. McGraw-Hill, New York, 1960.

On a theorem of V. M. Zolotarev. (Russian, English summary), *Teor. Veroyatnost. i Primenen. 9,* 99 (1964).

Asymptotically optimal tests for multinomial distributions. *Annals of Mathematical Statistics 36,* 369–400 (1965).

On probabilities of large deviations. "Proceedings of the Fifth Berkeley Symposium on Mathematical Statistics and Probability," Vol. I, pp. 203–219. University of California Press, 1967.

Some recent developments in nonparametric statistics. *Review of the International Statistical Institute 36,* 176–183 (1968).

(With Gordon Simons) Unbiased coin tossing with a biased coin. *Annals of Mathematical Statistics 41,* 341–352 (1970).

The L_1 norm of the approximation error in Bernstein-type Polynomials. *Journal of Approximation Theory 5,* 347–356 (1971).

On the centering of a simple linear rank statistic. *Annals of Statistics 1,* 54–66 (1973).

The L_1 norm of the approximation error for splines with equidistant knots. *Journal of Approximation Theory 11,* 176–193 (1974).

Some incomplete and boundedly complete families of distributions. *Annals of Statistics 5,* 278–291 (1977).

More on incomplete and boundedly complete families of distributions. "Statistical Decision Theory and Related Topics, II (Shanti S. Gupta and David S. Moore, eds.), pp. 157–164. Academic Press, New York, 1977.

SOME MEMORABLE INCIDENTS IN PROBABILISTIC/STATISTICAL STUDIES

Jerzy Neyman

Statistical Laboratory
University of California
Berkeley, California

I. INTRODUCTION

1. Congratulations to Professor Hoeffding. I am very grateful to Professor Chakravarti for his invitation to open the discussion at this Symposium intended to honor Professor Wassily Hoeffding. We met long ago and from the very beginning, it was a pleasure to find a marked similarity in our research interests. After the joint work with Egon S. Pearson [1933] concerned with power functions and later, after the development of the theory of confidence intervals [1937a], my research efforts focused on the deduction of variously defined "optimal" statistical methodologies [1959] that could be easily used in studies of natural phenomena. Against this, here is the title of Professor Hoeffding's paper: "Optimal nonparametric tests" [1951] he delivered at the Second Berkeley Symposium on Statistics and Probability held during the summer of 1950, more than a quarter of a century ago. Since that time our intellectual contacts continued, but our personal encounters were "like Victoria Regina: seldom, seldom in bloom."

Incidentally, the problem of the optimal non-parametric tests of composite statistical hypotheses is still "on the books."

II. THE CRAMÉR-HOEFFDING RESEARCH INCIDENT

2. The Harald Cramér Ground Breaking Paper of 1938. The mathematical tool most frequently used in the development of statistical

ISBN: 0-12-166650-6

methods is the Central Limit Theorem on probabilities, roughly as follows. Let $\{X_n\}$ be a sequence of random variables each having two moments, $EX_n = 0$ and $EX^2 = \sigma^2 < \infty$ and let

$$S_n = \sum_{i=1}^{n} X_i . \qquad (1)$$

Then, under certain conditions,

$$\lim_{n\to\sigma} P\{S_n \leqq t\sigma\sqrt{n}\} = \frac{1}{\sqrt{2\pi}} \int_{-\infty}^{t} e^{-u^2/2}\, du = \Phi(t) \qquad (2)$$

for any preassigned real number t. This theorem preoccupied mathematicians for a couple of centuries now [Loève, 1960]. The successive proofs given differ in the generality of the "certain conditions" just mentioned. This long duration of efforts to prove the validity of formula (2) resulted in the establishment of a "routine of thought." Whenever some particular problems of mathematical statistics involved the consideration of sums of random variables like (1), with the value of n considered "large," it became customary to presume that formula (2) gives a satisfactory approximation of the true distribution of S_n. The word "customary" is not adequate. The breaking of a "rountine of thought" stimulates opposition.

Among other things, the classical central limit theorem was used to compare the effectiveness of statistical tests. Here, the term Pitman asymptotic efficiency comes to my mind.

As described by Yu. V. Linnik [1961], the honor of breaking this firmly established routine of thought belongs to Harald Cramér. In 1938, just before the beginning of World War II, there appeared Cramér's paper [1938] offering the first solution to a novel question that Cramér dared to ask. Briefly, it is as follows.

With reference to formula (1) assume that all the variables of the sequence $\{X_n\}$ are mutually independent and identically distributed. Consider the probability

$$F_n(t_n) = P\{S_n \leqq t_n\sigma\sqrt{n}\}, \tag{3}$$

where t_n grows to infinity as n increased. Cramér's ground breaking question was about the asymptotic behavior of the ratio

$$\frac{1 - F_n(t_n)}{1 - \Phi(t_n)}, \tag{4}$$

depending on properties of the variables X_n and on the rate of increase of t_n. This paper generated a new chapter of probability theory, labeled "theory of large deviations" [Linnik, 1961]. Briefly and roughly, the important question was whether $1 - \Phi(t_n)$ can be considered as a satisfactory approximation of the probability that the sum S_n will exceed a limit proportional to $t_n\sqrt{n}$.

3. Professor Hoeffding's Initiative to Use the Novel Probabilistic Tool. While it is obvious that Cramér's limit theorem on large deviations must be a better tool for studying the asymptotic properties of statistical tests than is the classical central limit theorem, the disasters and the length of World War II were not conducive to the development of conceptual mathematical subdisciplines. In consequence, the relevance of the Cramér ground breaking work remained unnoticed for almost two decades. Here, a paper by Professor Hoeffding [1965] played a special role.

The title of this paper is:
"Asymptotically optimal tests for multinomial distribution."
Professor Hoeffding begins by formulating his own definition of asymptotic optimality and then states: "To attack these problems, the theory of probabilities of large deviations is needed." This is followed by proofs that, under specified conditions, certain familair tests (the likelihood ratio and the chi square tests) are asymptotically optimal in the sense of the new, call it, Hoeffding definition of optimality.

Professor Hoeffding's paper was presented at a meeting of the IMS and the discussion that followed is recorded in the Annals. It appeared that, even though Cramér's theorem on large deviations was familiar to several statisticians, including H. Chernoff, R.A. Wijsman and D.G. Chapman, Professor Hoeffding must be credited with the first serious effort to see what the novel probabilistic tool can contribute to the theory of asymptotic tests.

Incidentally, published in 1965, fourteen years ago, Hoeffding's paper continues to affect the thinking of this day. The following quote is from a paper published in the last issue of the Zeitschrift für Wahrscheinlichkeitstheorie und verwandete Gebiete [Berk and Jones, 1979]: "The first [lemma] is actually a special case of a theorem of Hoeffding (1965), Theorem 2.1."

My hearty compliments to Professor Hoeffding!

4. Reasons for Preferring the Theory of Large Deviations as a Tool for Studying Asymptotic Tests. The word "preferring" in the title of the present section emphasizes its subjective character . It has to do with the meaning I attach to the terms "errors of the first and second kinds" possible to commit in testing a statistical hypothesis.

As described in [1977a] in the course of an empirical study one is frequently faced with a two-decision problem. Depending upon the outcome of the statistical test used, one has to decide to go, say, either "right" or "left," and either decision can be erroneous. Depending upon personal attitudes, one of the two errors will be judged more important to avoid than the other. My definition is: the error that is more important to avoid is called the error of the "first kind." In consequence, when selecting a test to be used in a particular empirical study, my first concern is to make sure that the probability of committing an error of the first kind does not exceed a preassigned level α, now called "level of significance." Depending upon the subjective feeling of importance, the chosen level of significance may be $\alpha = 0.10$, or $\alpha = 0.05$ or $\alpha = 0.01$, etc.

When the problem of the desired level of significance is solved and if it can be ensured by any test of some determined class, the time comes to think of the less important error, the error of the "second kind," which means to determine the most powerful test within the class considered.

This is the background of my preference for the theory of large deviations as a tool in the theory of asymptotic tests as compared with the classical central limit theorem.

In an empirical study involving a two-decision problem, one is faced with some real life situation, with some hypothesis which can be true or false and with the degree of its falsehood measured by a parameter ζ, the value of which is unknown. The only thing that is under our control, at least to some extent, is the number n of observations that can be used to test the hypothesis that $\zeta = 0$. The all important question is whether this particular number n is large enough to achieve the chosen level of significance α. The answer depends on how close the ratio (4) is to unity, which is the subject of Cramér's theory of large deviations, including its modern descendants. The use of this theory does not violate the real life situation of the problem, with ζ having some unknown fixed value.

Now consider the asymptotic test possibilities offered by the classical central limit theorem on probabilities. As is well known, both the Pitman asymptotic efficiency theory and the theories of asymptotic tests developed by Cramér [1928] and by myself [1959] depend on visualizing that the real life problem, say, the problem of testing considered today, is a member of a hypothetical sequence with the fixed unknown ζ replaced by ζ_n, such that the product $\zeta_n\sqrt{n}$ is bounded away from zero and infinity, preferably tending to some known limit. This is something very different from and much less inspiring than the question of how close to unity is the value of (4).

III. TWO DIFFERENT STRATEGIES IN MATHEMATICAL STATISTICS

5. <u>A Curious Detail of the History of Statistical Tests</u>. The Cramér-Hoeffding research incident described in sections 2 and 3 illustrates a curious detail of the history of statistical tests, particularly of the early history. The customary strategy is composed of two consecutive steps. (i) A statistician concerned with some empirical domain proposes a testing procedure suggested by his intuition. Then, (ii) an effort is made to investigate the properties of this procedure, occasionally leading to the conclusion that it is in some sense "optimal." Examples of this sequence (i)-(ii) are countless.

The first test procedure, still in very frequent use, is the chi square test introduced by Karl Pearson in 1900. It was one of the subjects studied in the Hoeffding paper just discussed. The other test discussed in the same Hoeffding paper is the likelihood ratio test. As stated by Professor Hoeffding, the likelihood ratio criterion was suggested by E.S.P. and myself in 1928. However, this suggestion was made on intuitive grounds. The criterion suggested did not result from a search for a procedure satisfying a defined concept of optimality. The intuitive background of the likelihood ratio test was simply as follows: if among the contemplated admissible hypotheses there are some that ascribe to the facts observed probabilities much larger than that ascribed by the hypothesis tested, then it appears "reasonable" to reject that hypothesis.

As another example, I wish to mention a test criterion competitive to the chi square, first suggested by Harald Cramér [1928] and somewhat later also advanced by Richard von Mises [1931].

The alternative philosophy, or strategy, is just the opposite to the sequence (i) and (ii). When one has to deal with an empirical domain of study and one feels in need of a statistical procedure, it seems natural to visualize the properties that this procedure should have to deserve the description "optimal." Naturally, such concept of optimality <u>can depend</u> upon the domain of empirical study and it <u>must depend</u> on the subjective preferences of its author. However, once the optimality is defined, the mathematical problem occurs: to find the "optimal," if such exists. On occasion one finds that the initially defined optimal procedures do not exist. Too bad! Then one has to look for a "compromise optimality," etc. One example is the concept of "unbiased most powerful tests" [Neyman and Pearson, 1936]. Here, the word <u>unbiased</u> marks the compromise optimality. In the case considered, the "uniformly" most powerful test does not exist.

IV. THE YULE-PÓLYA RESEARCH INCIDENT: (i) MECHANISM OF A NATURAL PHENOMENON, AND (ii) NON-IDENTIFIABILITY

6. <u>My Contacts with George Udny Yule</u>. During my four year long activities at the Department of Statistics, University College, London (1934-1938), I had the privilege of meeting quite a few outstanding scholars. This included G. U. Yule for whom I developed great respect and warm feelings.

The studies of Yule that attracted my particular attention were performed jointly with M. Greenwood [1920]. Subsequently, a related paper was published by E. M. Newbold [1928]. My preferred way of describing these studies is as follows: They are concerned with the chance mechanism operating in real life, the mechanism that determines the distribution of an observable random variable X. If this mechanism is understood, it could be used to solve an important practical problem.

The particular random variable X of the Greenwood-Yule-Newbold studies was the number of accidents per unit of time, per bus driver in London. The important practical problem considered was the means to diminish the frequency of accidents involving the buses. Exactly similar problems are important in the present epoch, even though the actual domain of study can be very different. One example is the question: how can one diminish the frequency of deaths from cancer?

The problem of accidents was studied in our Stat. Lab. in the early 1950's. Here Professor Grace E. Bates played an important role [Bates and Neyman, 1952a, 1952b].

The first of these papers is dedicated to the memory of George Udny Yule and is preceded by a one page biographical sketch. It includes the following passage: "In 1931 Yule felt that he was too old to hold the position of Reader at Cambridge University and retired. At the same time he felt young enough to learn to fly. Accordingly, he went through the intricacies of training, got a pilot's license and bought a plane. Unfortunately, a heart attack cut short both the flying and, to a considerable degree, his scholarly work."

It happened that my personal contacts with Yule were very limited. They occurred during the period when he was recovering from his heart attack. However, these contacts affected my thinking. In particular, they contributed to the formulation of my paper of 1937[b].

The attempts to decrease the frequency of accidents taking into account the "human factors," mentioned in the title of Miss Newbold's report, are connected with the concept now called "accident proneness." There is little doubt that particular individuals do differ in their proneness to accidents of some specified categories. However, the details of this variability are not clear and here empirical studies are important. During our studies in the early 1950's our thinking was affected by two contrasting hypothetical mechanisms. One of them is the Greenwood-Yule-Newbold (GYN, for short) hypothetical mechanism, the properties of which can be summarized as the "mixture - no contagion - no time effect" mechanism. The other hypothetical mechánism, implied by studies of George Pólya [1930], was just the contrary: "identity of individuals, contagion and time effect."

To be more specific: the GYN mechanism presupposed that the number of accident incurred by a particular individual per unit of time, such as a year, is a Poisson variable with a fixed expectation λ, representing this individual's personal accident proneness, which remains unchanged throughout his active life (= "no time effect"). Another basic assumption is that the value of λ varies from one individual to the next (= "mixture"). More particularly, the assumption was adopted that the variation of λ within a relevent population, such as the population of actual or potential bus drivers in London, can be adequately represented by a gamma distribution.

Starting with these basic assumptions it was easy to deduce that the number of accidents per year incurred by individual bus drivers must have a negative binomial distribution. Actually, using the data on accidents involving bus drivers it was found that this distribution could be well fitted by a negative binomial so that the GYN mechanism (or shall we call it "model?") appeared to have been "confirmed."

Everything appeared nice and smooth until the Pólya "model" was examined. As described above, this model denied the existence of a "mixture." The basic assumption was that all individuals forming the population of actual or potential employees in a particular industry were "born equal." However, it was assumed that the number of accidents in a time interval [t, t+h), where h is a small positive number, depends upon the number of accidents incurred before time t (= "contagion"). Also, there was the assumption that, as the duration of employment increases, the experience gained may diminish the individual's accident proneness (= "time effect").

Using these specific assumptions suggested by the famous Pólya paper of 1930, it was easy to calculate the distribution of the number of accidents per year in a population comparable to that of the London bus drivers. Because of the contrast between the two hypothetical mechanisms, the GYN and the Pólya mechanisms, the expectation was that the two distributions would be very different. If this happened, then the empirical data, such as the data resulting from Miss Newbold's study of the London bus drivers could be used to resolve questions like that in the title of our study [1952b]: "true or false contagion?"

When the easy calculations of the relevent probability generating function were performed, Dr. Bates and I experienced a little shock: with reference to a single observational period, such as a year, the Pólya "no mixture - contagion - time effect" model implied that the distribution of the number of accidents per driver must be a negative binomial, coinciding with that implied by the Greenwood-Yule-Newbold model! This finding brought to our minds several ideas that appear important to this day. One is the concept of non-identifiability. The other related idea is that the problem of validation of a hypothetical mechanisms of a natural phenomenon deserves a serious effort. One hopeful possibility is that the non-identifiability of some two (or more) hypothetical mechanisms, the non-identifiability with respect to the distribution of a specific single random variable X, may disappear just as soon as one supplements X by some other appropriately selected variables, say $X_1, X_2, \ldots X_s$.

The second of our joint papers considers a number of not too difficult empirical studies capable of providing a definitive answer to the all important question about the reality of "contagion" in accidents. E.g., the identifiability can be achieved by counting accidents of each driver not just in one particular year (say X_1 of them), but also those incurred during the following year, say X_2 of them, etc. See Grace E. Bates [1955].

This section concludes my description of the Yule-Pólya problem as it came to my attention with reference to industrial accidents: what is the governing chance mechanism? Without much risk of exaggeration one may assert that this type of problem is encountered in every serious study of a complex natural phenomenon. In cosmology: what is the chance mechanism governing the dispersal of clusters of galaxies? How can one verify any relevent hypothesis? In public health: what is the mechanism behind the observed geographic variability in the incidence of cancer? Through what experiments and with what statistical methodology can one gain reliable information? In weather modification experiments: what are the processes in the atmosphere that follow "cloud seeding?" What statistical methodology is likely to provide the desired information through the analysis of the many completed experiments?

Here, a remark on terminology seems in order. It seems to me that the common use of the term "model" deserves a modification or restriction. My preference would be to restrict the use of this term to sets of (customarily) qualitative assumptions advanced to explain a natural phenomenon. One example is the GYN model suggested to explain the notorious driver to driver variability in the number of accidents per year, the "mixture - no contagion - no time effect" model. The same applies to the Pólya "no mixture - contagion - time effect" model. This use of the term "model" appears quite different from the designation of a mathematical formula that fits the observations. One frequently encountered example is the phrase "linear model," etc.

Discussions of the Yule-Pólya dilemma relating to the problem of public health will be found in the next chapter.

V. SOME PRESENT DAY RECURRENCES OF THE YULE-PÓLYA DILEMMA

7. Public Health Policy and Basic Research. The importance and the difficulty of the present day public health problems overshadow those of industrial accidents symbolized by the names of Yule and Pólya. However, the broadly understood research problems remain similar.

One of the typical contemporary public health problems is concerned with the hazards from electricity producing plants [1977b], briefly as follows. A locality L, marked by a rapidly growing population, is in need of a new electricity producing plant. This may be either a nuclear facility or a fossil fuel burning unit and the choice is up to some decision making authorities. Among other things, the choice must be made taking into account some public health questions. Whatever type of plant is constructed, it will contribute to the local pollution in its own way. The important questions are: how many more cancer cases, heart attacks, etc. are to be expected in this locality L as a result of the predictable extra pollution from the normal operation of the novel electric generator? How can one answer this question reliably?

The reliability of the answer depends upon the understanding of two different mechanisms. One mechanism is concerned with the happenings in experimental animals, mice, dogs, etc., subjected to a specified change in the environmental pollution. The other important mechanism is that of the dependence of the effects of the first mechanism on the identity of the species concerned, whether mouse, or rat, or dog, or man. Obviously, the complexity of the problem is tremendous. It splits itself into a number of subproblems. In the next section, we shall consider one of these subproblems. It involves the ubiquitous phenomenon of non-identifiability.

8. Typical "Survival Experiment" and the Methodology of "Potential Survival Times." The customary source of information on the happenings in the experimental animals, say mice, exposed to some "agents" studied is a "survival experiment." There are two substantial groups of mice, one labeled "experimental" and the other "controls." The experimental mice are exposed to the agents studied and the controls are not. When a mouse of either group dies, its body is subjected to a pathological study and an effort is made to determine the cause of its death. With a degree of oversimplification, it is postulated that there is a somewhat limited number of possible causes of death, say K of them. The problem studied is that of the difference in death rates from the different causes among the experimental and the control mice. This is only a rough description of the problem. One of the difficulties that became obvious on closer examination is due to the omni-present phenomenon of "competing risks." One illustrative example is as follows.

All of us alive today are exposed to a variety of risks of death, including street traffic and cancer. If I am run over and killed by a car tonight, it would be impossible for me to die later from cancer and, in due course, this would affect the published death rates from cancer. In consequence, the numerical results of a survival experiment with mice do not characterize "net rates" of deaths from the various causes of death studied but only the "crude rates." These crude rates corresponding to the different causes (or "risks") studied characterize not only the intensities of particular risks, but they also reflect the combined property of all of them that is due to competition. Now, let us visualize the results of a completed survival experiment after all the mice, say of the experimental group, have died.

Table 1 illustrates the obtainable results.

TABLE I. Illustration of the results of a survival experiment

Cause of Death	Survival Times of Particular Mice
C_1	$t_{11} \leqq t_{12} \leqq t_{13} \dots \leqq t_{1n_1}$
C_2	$t_{21} \leqq t_{22} \leqq t_{23} \dots \leqq t_{2n_2}$
...	....
...	....
C_K	$t_{K1} \leqq t_{K2} \leqq \dots\dots \leqq t_{Kn_K}$

The first column of Table I enumerates all the K causes of death. The wide second column gives the corresponding consecutive survival times of mice that died from the particular causes. Thus, for example, the symbol t_{11} stands for the time of the first recorded death from cause C_1. Similarly, the last symbol in the same line, namely t_{1n_1} represents the time of death of the last mouse that died from the same cause C_1, etc. Here, then, the subscripts $n_1, n_2, \dots, n_K$ denote the numbers of mice that died from causes $C_1, C_2, \dots C_K$, respectively. Naturally, these numbers $n_1, n_2, \dots, n_K$ will not be all equal and their variability will reflect both the severity of particular causes and their competition. The reader will have no difficulty in visualizing an exactly similar table compiled for the control mice. These two tables would then be ready for the evaluation of the effects of the agents studied on the survival experience of the mice.

Having in one's mind the problem of a new electric generator in locality L, one might think of the question: how many more deaths from cancer (perhaps cause C_1) should one expect among mice if the "agents" studied included irradiation? What about the methodology of evaluating the experiment that could answer reliably a question of this kind?

One of the methodologies used is that, based on the concept of "potential survival times." For an experimental animal exposed to K possible risks (or causes) of death, the term i-th potential survival time designates a random variable Y_i supposed to represent the age at death of this animal in the hypothetical condition in which C_i is the only possible cause of death. The probability that Y_i will exceed a preassigned value t is called the "net survival probability."

Unfortunately, while a survival experiment can be conducted to investigate a great variety of different "agents," the resulting "causes" of death are not under control of the experimentor. Thus, no direct empirical counterpart of the net survival probability can be available. All that the results of a survival experiment illustrated in Table I can provide is the empirical counterparts of the so-called "crude survival probabilities." For the i-th cause the crude probability of surviving up to time t, say $Q_i(t)$ is the probability that $Y_i = \min(Y_1, Y_2, \ldots Y_K)$ and that $Y_i > t$. Here, then, the question arises whether a statistical methodology could be developed to use the crude survival data as in Table I, perhaps somehow supplemented, in order to estimate the net survival probabilities.

An interestingly described by David [1974], the competing risk phenomenon occurs not only in problems of public health but also in problems of technological reliability. Here, the most attractive presumption supplementing the data of a survival experiment is the assumption that the potential survival times Y_i are mutually independent. However, the hypothesis of independence cannot be tested using the data of a survival experiment and the publications of Tsiatis [1975] and of Peterson [1976] document the presence of non-identifiability. The crude survival probabilities are consistent with an infinity of systems of widely different net survival probabilities. The conclusion is that the survival experiment of the type described is too simplistic to provide all the valuable information for studies of problems of health.

9. Survival Experiments with Serial Sacrifice. The "serial sacrifice" methodology [Upton, 1969] represents a very important advance in the health related experimentation. Rather than focus on the diagnosed "causes" of death of the experimental animals, the serial sacrifice experimentation deals with what I like to call "elementary pathological states," say S_1, S_2, ... S_K. For example S_1 may stand for thymic lymphoma (a cancer), S_2 for reticulum cell sercoma, another cancer, etc. At selected times, say t_1, t_2, ... samples of mice alive at these times are killed (= "sacrifice") and their bodies are subjected to a pathological analysis. The result of such analysis for a particular mouse may be that, at the time of its sacrifice, it was affected by, say, three elementary pathological states, S_4, S_5, S_6, and no others.

The above methodology provides empirical counterparts to the following type of questions: how frequently the mice alive at the preassigned times t_1, t_2, ... are affected by this or that combination of pathological states? Combined with similar data for mice that died on their own (not through "sacrifice") the amount of information from a serial sacrifice experiment is very much richer than from the "typical" survival experiment illustrated in Table I. Also, there is an important difference in the nature of the information.

Here, I wish to call the reader's attention to the analogy between the serial sacrifice vs. "typical" survival experiment situation, on the one hand, and the multiple periods of counting accidents vs. just one such period, on the other. As discussed in Section 6, the non-identifiability of two contrasting mechanisms of accident proneness was due to the insufficiency of observational data: numbers of accidents incurred during a single year. The counts of accidents incurred by each driver the following year made the non-identifiability disappear. It is this analogy that is symbolized by reference to the "Yule-Pólya dilemma" in the title of the present Chapter V.

I learned about the serial sacrifice design during a visit to the Oak Ridge National Laboratory and, particularly, through conservations with Dr. John B. Storer. At the time Dr. Storer was in charge of the continuing experiment set up by Upton. Later, we had the pleasure of Dr. Storer's visit to Berkeley. Also, we received from him a substantial sample of data from the experiment in question. In these data, the total number of elementary pathological states was eight. The further difference with the "typical" survival experiment was that there was no "causes" of death indicated.

While all human determinations are subject to error, the determination of particular pathological states is comparable to chemical analyses and represents an effort at objectivity. On the other hand, the diagnosis of a "cause" of death is a conclusion likely to be affected by subjective attitudes of the pathologists.

10. Another Shock of Non-Identifiability. As mentioned in Section 6, the finding of non-identifiability affecting the study of accident proneness caused Dr. Bates and myself to experience a shock. Here, I have to admit a somewhat explosive feeling of enthusiasm I felt when contemplating the experimental results obtainable through serial

sacrifice experiment. I rather felt that these results, without any additional observations, provide data for the study of a stochastic process representing the natural succession of life and death events: birth at time zero, followed by first illness at age t_1, then by recovery at time t_2, etc. etc., and finally death at some observable time. Because the domain of stochastic processes is now well developed, I expected that a statistical methodology could be discovered to use the serial sacrifice data in order to estimate the mechanism of treatment effects in mice contemplated, perhaps, as a realization of a finite states Markov chain, with all the transition probabilities possible to estimate. Due to the work of Clifford [1977], I experienced a shock. Even with some over-simplifying assumptions (denying the possibility of "recovery," etc.) a discrete time Markov chain model proved to be unidentifiable with respect to the data of a serial sacrifice experiment! The details are described in the analysis of Storer's data performed with Clifford's active participation [Berlin et al, 1979].

While the serial sacrifice data provide answers to the questions "how frequently mice sacrificied at age t are affected by a stated combination of pathological states," the missing information relates to mice alive at age t and having at that age a stated pathological combination. During the subsequent unit of time, say during the next 100 days, the health state of these mice can change in many different ways: recovery from some illnesses, contracting some others, etc. With the present design of serial sacrifice experiments there is no information on the frequency of such transitions. The tantalizing question is whether some not too difficult modification of the methodology could provide information to fill in the now existing gaps. The way of discovering such effective modifications requires a reasonably close cooperation between an intensely interested statistician and an equally intensely interested experimenting biologist. The questions to resolve are of the following type: could the analysis of urine of a mouse provide enough information on its health state? Could the analysis of a blood sample be sufficient? However, can this sample of blood be taken without altering the contemporary transition probabilities of the mouse, i.e., without hurting the mouse? Who knows? However, unless one tries, one can hardly hope to succeed.

VI. EFFORT AT AN "OPTIMAL" COMPETITOR TO K.P.'S χ^2 TEST FOR GOODNESS OF FIT

11. Introductory Remarks. This chapter is intended to illustrate my preferred strategy of studying or of developing statistical tests: begin by defining the optimal performance of the test, and then try to deduce the desired criterion. As indicated in the title of the chapter, the example chosen for illustration is the Karl Pearson's test "for goodness of fit" symbolized by χ^2.

As is well known, the χ^2 test is now being used for a variety of purposes, such as contingency tables, etc. In these circumstances, I wish to emphasize the limited scope of the following discussion: it is concerned with the problem of "goodness of fit" as contemplated in olden days by K.P. My actual effort to formulate the problem and to solve it was published in 1937[b]. It is limited to the case of a "simple hypothesis," that is, to the case in which the problem is to decide whether a completely specified probability density, say $p_X(x)$ fits the empirical distribution of an observable random variable X. Another limitation consists in the assumption that the number N of observed values of X is "large." The problem of extending the methodology to the case of composite parametric hypotheses has been treated by Javitz [1975].

12. Criticism of the K.P.'s χ^2 Test for Goodness of Fit. An effort at an "optimal" competitor of an existing test intended for use in some specified conditions must begin by the unavoidably subjective criticism of the original test. The well known procedure of the χ^2 test for goodness of fit begins by dividing the range of variation of the observable X into a certain number, say s, of "cells," with boundaries

$$a_0 < a_1 < a_2 \dots < a_s, \tag{5}$$

where a_0 may mean $-\infty$ and a_s may be $+\infty$. Next, the probability density $p_X(x)$ is used to compute the expected number of independent observations, say n_i, falling into the i-th cell for $i = 1, 2, \dots s$. Let m_i denote the actual number out of the total N observations that fall into the same i-th cell. Then, K.P.'s test criterion for goodness of fit is given by

$$\chi^2 = \sum_{i=1}^{s} \frac{(m_i - n_i)^2}{n_i} \tag{6}$$

The fit is considered "bad" if the calculated χ^2 exceed the tabled limit corresponding to the chosen level of significance. Otherwise, the fit is considered "good."

My own subjective criticism of the test includes the fact that the value of the criterion (6) does not depend on the order of positive and negative differences $(m_i - n_i)$. The extreme example is represented by the following possibilities. In one case, the signs of the consecutive differences $m_i - n_i$ and $m_{i+1} - n_{i+1}$ are not the same. In the other case one can observe a substantial number of consecutive differences $m_i - n_i$ that are all negative while all the others are positive. While these two possibilities are consistent with the same value of the criterion (6), my intuitive feeling is that in the second case the "goodness of fit" is subject to a rather strong doubt, irrespective of the actual computed value of (6), even if it happens to be small.

13. <u>"Smooth Test" for Goodness of Fit</u>. The first step in the deduction of the "smooth test" intended as an "optimal" competitor to χ^2, consisted in standardizing the analytical developments. Rather than consider the great variety of distributions $p_X(x)$ that may come under consideration, I proposed to replace the observable X by its function Y defined by the relation

$$y = \int_{-\infty}^{x} p_X(x)dx \tag{7}$$

where y and x designate particular values of the two random variables. As it is easy to check, the range of variation of Y is from zero to unity, with its probability density

$$p_Y(y) = 1, \tag{8}$$

this, irrespective of the distribution of X.

As contemplated by Karl Pearson, the background of the problem of goodness of fit admits the possibility that the specified density of $p_X(x)$ may not correspond to reality. However, there are no general indications as to what the alternatives might be. In my attempt to deduce an optimal competitor to the chi square test, I contemplated the set of alternatives vaguely described as "smooth."

In terms of the variable Y, with its range of variation limited to the interval (0,1), where its density is equal to unity, the contemplated "smooth" alternatives are those with densities the logarithms of which are polynomials of orders 1, 2, ... K. The theory published in 1937 develops an asymptotic version of optimal unbiased type C tests of orders $K = 1, 2, \ldots$ with K denoting the order of polynomial used.

The study of asymptotic power of these tests indicates that, generally, adequate results could be obtained with K not exceeding 4. The tests so deduced are not open to the criticism of the original test for goodness of fit indicated above.

In recent times quite a few non-parametric tests for goodness of fit have been considered with emphasis on their robustness. It would be interesting to use the Monte Carlo methodology to compare the performance of these tests with that of the smooth test of a limited order $K \leqq 4$.

ACKNOWLEDGMENTS

This paper was prepared using the facilities of the Statistical Laboratory with partial support from the Office of Naval Research (ONR N00014 75 C 0159), the Department of the Army (Grant DA AG 29 76 G 0167), and the National Institute of Environmental Health Sciences (2 R01 ES01299-16). The opinions expressed are those of the author.

REFERENCES

Bates, Grace E. (1955). "Joint distribution of time intervals for the occurrence of successive accidents in a generalized Pólya scheme," Ann. Math. Stat., Vol. 21, pp. 705-720.

Bates, Grace E. and Neyman, J. (1952a). "Contribution to the theory of accident proneness, I. An optimistic model of correlation between light and severe accidents," Univ. of Calif. Publ. in Stat., Vol. I, pp. 215-254.

Bates, Grace, E. and Neyman, J. (1952b), "Contribution to the theory of accident proneness, II. True or false contagion," Univ. of Calif. Publ. in Stat., Vol. I, pp. 255-276.

Berk, Robert H. and Jones, Douglas M. (1979). "Goodness-of-Fit Test Statistics that Dominate the Kolmogorov Statistics," Zeitschrift für Wahrscheinlichkeitstheorie und verwandte Gebiete, Vol. 47, pp. 47-59.

Berlin, Bengt, Brodsky, Joel and Clifford, Peter (1979). "Testing Disease Dependence in Survival Experiments with Serial Sacrifice," JASA, Vol. 74, pp. 5-14.

Cramér, Harald (1928). "On the composition of elementary errors," Skandinavisk Aktuarietidskrift, Vol. 11, pp. 13-74 and 141-180.

Cramér, Harald (1938). "Sur un nouveau théorème-limite de la théorie des probabilités," Actualités Sci. Ind. No. 736, pp. 5-23.

David, H.A. (1974). "Parametric approaches to the theory of competing risks," Reliability and Biometry, Statistical Analysis of Lifelength (F. Proschan and R.J. Serfling, eds.), SIAM, Philadelphia, pp. 275-290.

Greenwood, M. and Yule, G.U. (1920). "An inquiry into the nature of frequency distributions ... with particular reference to ... repeated accidents," J. Roy. Stat. Soc., Vol. 83, pp. 255-279.

Hoeffding, Wassily (1951)." 'Optimum' Nonparametric Tests," Proc. Second Berkeley Symp. Math. Stat. and Prob., Univ. of California Press, Berkeley, CA, pp. 33-92.

Hoeffding, Wassily (1965). "Asymptotically optimal tests of multinomial distribution," Annals of Math. Stat., Vol. 36, pp. 369-401.

Javitz, Harold S. (1975). "Generalized smooth tests of goodness of fit, independence, and equality of distributions," Unpublished doctoral dissertation, Univ. of Calif., Berkeley.

Linnik, Yu. V. (1961). "On the probability of large deviations for the sums of independent variables," Proc. Fourth Berkeley Symp. Math. Stat. and Prob., Vol. II, Univ. of Calif. Press, Berkeley, CA, pp. 289-306.

Loève, Michel (1960). Probability Theory, Van Nostrand, p. 268.

Mises, Richard von (1931). Wahrscheinlichkeitsrechnung, Leipzig u. Wien, pp. 316-335.

Newbold, E. M. (1928). "A contribution to the study of human factors in the causation of accidents," Industr. Health Res. Board Report No.34, London, H.M. Stationary Office.

Neyman, J. (1937a). "Outline of a theory of statistical estimation based on the classical theory of probability," Philos. Trans. Roy. Soc. of London, Ser. A., Vol. 236, pp. 333-380.

Neyman, J. (1937b). "Smooth test for goodness of fit," Skandinavisk Aktuarietidsckrit, Vol. 20, pp. 149-199.

Neyman, J. (1959). "Optimal asymptotic tests of composite statistical hypotheses ," Probability and Statistics (The Harald Cramér Volume), (U. Grenander, ed.), Almquist and Wiksells, Uppsala, Sweden, pp. 213-234.

Neyman, J. (1977a). "Frequentist probability and frequentist statistics," Synthese, Vol. 36, pp. 97-131.

Neyman, J. (1977b). "Public health hazards from electricity-producing plants," Science, Vol. 195, pp. 754-758.

Neyman, J. and Pearson, E.S. (1933). "On the problem of the most efficient tests of statistical hypotheses," Philos. Trans. Roy. Soc. of London, Ser. A., Vol. 231, pp. 289-337.

Neyman, J. and Pearson, E.S. (1936). "Contributions to the theory of testing statistical hypotheses (1) Unbiased critical regions of Type A and Type A_1," Stat. Res. Memoirs, Vol. 1, pp. 1-37.

Pearson, Karl (1900). "On the criterion that a given system of deviations from the probable in the case of a correlated system of variables is such that it can be reasonably supposed to have arisen from random sampling," Phil. Mag. and J. of Sci., Vol. 50, pp. 157-175.

Peterson, A. V. (1976). "Bounds for a joint distribution function with fixed sub-distribution functions: application to competing risks," Proc. Natl. Acad. Sci., USA, Vol. 73, pp. 11-13.

Polya, G. (1930). "Sur quelques points de la théorie des probabilités," Ann. de l'Institut Henri Poincaré., Vol. 72, No. 1, pp. 117-161.

Tsiatis, A. (1975). "A nonidentifiability aspect of the problem of competing risks," Proc. Nat'l Acad. of Sci., Vol. 72, No. 1, pp. 20-22.

Upton, A.C. et al (1969). Radiation Induced Cancer, International Atomic Energy Agency, Vienna, p. 425.

VII. SERIAL SACRIFICE EXPERIMENTS
AN ADDENDUM

1. Why an Addendum? Survival experiments, including those with serial sacrifice, are discussed in the main body of this article in Chapter V. The title of this chapter is "Some Present Day Recurrences of the Yule-Pólya Dilemma." The dilemma in question consists in the non-identifiability of two contrasting mechanisms of a phenomenon which it is important to understand in order to be able to solve a particular practical problem. My motivation for illustrating this kind of dilemma in modern times using the survival experiments has been the importance of the policy question: what can one do to diminish the frequency of deaths from cancer? Obviously, in order to be able to answer this question, one has to understand the mechanism of carcinogenesis apparently caused by a number of different agents, including irradiation.

As described in Section 9 of the main body of the present paper, my familiarity with the serial sacrifice methodology stems primarily from contacts I had with Dr. John B. Storer who, at the time, was in charge of a continuing serial sacrifice experiment set up by Dr. Arthur C.

Upton. It is here that Dr. Peter Clifford discovered the ubiquitous phenomenon of non-identifiability. The present Addendum is motivated primarily by the recent appearance of two articles discussing this same experiment. Regretfully, it was only a few days ago that I became familiar with these articles. My additional motivation is the idea of an interesting mathematical problem that relates to the Clifford non-identifiability.

2. The Upton-Storer Experiment. The two articles just mentioned are due to the same two authors, Toby J. Mitchell and Bruce W. Turnbull. The two papers are published in the prestigious journal Biometrics, one in 1978 and the other in 1979. The titles of the two papers indicate related contents:

"Exploratory Analysis of Disease Prevalence Data From Survival/Sacrifice Experiments:

and

"Log-Linear Models in the Analysis of Disease Prevalence Data From Survival/Sacrifice Experiments,"

respectively. The summary of the more recent paper specifies its relation to the earlier one. In part, this summary reads as follows:

> This paper considers the problem of analyzing disease prevalence data from survival experiments in which there may also be some serial sacrifice. The assumptions needed for "standard" analyses are reviewed in the context of a general model recently proposed by the authors. This model is then reparametrized in log-linear form, and a generalized EM algorithm is utilized to obtain maximum likelihood estimates of the parameters for a broad class of unsaturated models. Tests ... are proposed to investigate the effects of

> treatment, time, and the presence of other diseases on the prevalences and LETHALITIES of specific diseases of interest...." [Emphasis added.]

The above passage, mentioning reparametrizing, stimulates interest in "context" of the earlier paper. My personal interest focuses on the emphasized word "lethalities" of the specific diseases of interest.

The earlier paper of the two authors reflects their broad familiarity with the relevant literature. This includes the non-identifiability results of Tsiatis, of Clifford and of the team of Berlin, Brodsky and Clifford. The latter paper is mentioned as an unpublished manuscript. The reference includes the book of C.L. Chiang and also a paper by B. Efron expected to appear in 1978 under the colorful title "Bootstrap methods: another look at the jackknife." The method of analysis, under the title "Model," is explained on pp. 556-557.

> As in a life table analysis, the time scale, which refers to the age of the animal, is partitioned into $m + 1$ intervals by the points t_1, t_2, ... t_m. Each interval will be considered separately ....
>
> Similar to an illness-death model (Chiang 1968) we assume that there are $K(> 1)$ "illness states," each corresponding to one of the 2^I possible combinations of I diseases; one of the states is the absence of all I diseases. We let p_i be the probability that an individual in the control group, alive at the beginning of the interval, is in illness state i at the time; P_1 is defined similarly for the treated group. We also define q_i as the probability that an individual in state i at the beginning of the interval dies in the interval ....

Implicit in this model are two basic approximating assumptions:

A1. No transition between illness states occurs during the interval.

A2. The lethalities (q_1's) are the same for both treated and control groups, i.e., the probability of death depends on group membership only through the illness states.

3. Comments on the "Two Basic Approximating Assumptions." The approximating assumption A1 does not seem to me realistic. The duration of time intervals considered by the two authors is 50 days. This is a long period. As our personal experiences indicate, during a period of fifty days one can easily contract an illness, possibly have some complications, and also recover from both, from the illness and from the subsequent complications, etc. This is certainly true with humans and is also likely to be true with animals used for the experiments. Rather than assume that all the fifty days' transition probabilities are zero, a more realistic "approximating assumption" might be that during the fifty days these transition probabilities do not change very much and might be considered constant.

Because of the above considerations, the further discussion in this Addendum is independent of the "basic approximating assumption" A1.

The reason for my next comment on the Turnbull/Mitchell methodology can be symbolized by the difference in the terminology we use. In discussing the Upton-Storer experiment the two authors speak of the "DISEASE prevalences" that can be estimated. This contrasts with my term "elementary pathological states" (see Section 9). This latter term applies to experimental data exemplified by the following table reproduced from Neyman and Scott (1967), p. 758.

COUNTS OF CELLS, OF HYPERPLASTIC FOCI, AND OF TUMORS IN LUNGS OF MICE AFTER SHIMKIN AND POLISSAR [12].

	Estimated Mean Number of:			
Days after Urethane	Cells per Square (106.3 sq. micra)	Presumed First Mutants per Square	Foci per Lung	Tumors per Lung
0	0.73	0.00	-	-
1	0.85	0.12	-	-
3	0.92	0.19	-	-
7	1.11	0.38	-	-
14	1.02	0.29	294	-
21	1.35	0.62	450	-
28	1.57	0.84	390	15.5
38	-	-	610	-
49	1.33	0.60	450	37.3
84	1.20	0.47	260	34.8
105	-	-	200	35.2
133	-	-	83	35.7

The table gives day-to-day "prevalences" of a substantial number of "elementary pathological states" such as somehow modified cells, of "hyperplastic foci" and of "tumors per lung" of the experimental mice. For a statistician it seems possible that the availability of some such details of serial sacrifice experiments would contribute to the understanding of the mechanism of carcinogenesis and may alleviate the difficulty of non-identifiability.

My next comment on the same paper of the two authors refers to the apparent inconsistency between the last line of the Summary (p. 555) and the description of the actual experiment studied. The Summary mentions "the effects of low-level radiation on laboratory mice." This compares with the following descriptions on p. 561:

> The data we consider here consists of two groups of ...
> mice, a treated group which received 300 rads of gamma

radiation delivered at 45 rads/min. ... and a control group which received none.

My point is that the dose-rate of 45 rads per minute is rather high and the description of the experiment in terms of "low-level" radiation may mislead some readers of Biometrics. The same gamma radiation administered at the rate of, say, one rad per day (not per minute) would be closer to what one might call "low level" and its carcinogenic effects are likely to be quite different.

4. An interesting mathematical problem. The interesting mathematical problem I have in mind is connected with the "basic approximating assumption A2."

My point is that the description of A2 as a basic APPROXIMATING assumption is really a misnomer. Rather, A2 is an unsupported presumption regarding the effect of the treatment which is one of the subjects of study. Specifically, the assumption A2 is that, for mice of a given age affected by a specified combination of diseases the probability of death during the next 50 days is the same for the control and for the irradiated groups. But is it not possible that the intensities of the diseases among the controls is very different from those among the irradiated mice? With the non-identifiability proved by Clifford, I am interested in the mathematical methodology for calculating how large could be the differences in the probabilities of death. Naturally, these differences could (and very likely WOULD) depend on the conditions considered: age and health of the mice, the total dose and the dose-rate of irradiation. In some conditions the changes could be from almost zero for control mice to almost unity for those irradiated. In some other conditions, the change could be, say from 0.6 to 0.5, etc.

The problem I have in mind is analogous to that solved by A.V. Peterson (1976). His paper is briefly mentioned in Section 8 of the main body of this article. This section is concerned with the methodology of "potential survival times." Here, Tsiatis proved non-identifiability, while Peterson found the "bounds" of this non-identifiability. What are the "bounds" of the Clifford non-identifiability?

5. Appendix by Peter Clifford (Mathematical Institute, Oxford University, England.) Professor Neyman has asked how bounds may be obtained for the non-identifiability inherent in the fitting of a Markov model of illness and death to data from a serial sacrifice/survival experiment. In recent conversations with him, I was informed that the problem was probably difficult. Although I may be accused of considering a favorable interpretation of the question, I have to disagree. The solution involves a certain amount of Graph theory and familiarity with the techniques of Linear Programming. Here, I will give an outline of the results obtainable and I hope to publish a more detailed exposition elsewhere.

On examination, after death, an animal in a serial sacrifice/survival experiment may be observed to be in one of a finite number of pathological states. During the lifetime of the animal it is plausible that certain transitions between states may occur. Taken together these assumptions specify a certain directed graph, the transition graph. To each plausible transition there corresponds an age dependent probabilistic intensity of transition. It is possible to show that the non-identifiability of these parameters results from the presence of cycles in the transition graph. The worst case is when there is a directed cycle that is a path, following plausible transitions, starting and ending at the same state. In this case the indistinguishable set of parameters can be shown to be unbounded. At the other extreme, if there are no cycles then all parameters are identifiable. When there are cycles but no directed cycles the set of indistinguishable parameters is convex and bounded. Furthermore it is possible to investigate the extreme points of this set by Linear Programming. In a variety of simple cases explicit bounds can be given; in more complicated cases efficient computer algorithms may be used to investigate the set of indistinguishable parameters.

Thus bounds may be obtained for the interstate transition intensities, but the survival probabilities and lethalities depend on these intensities in a complicated manner and it appears to be difficult to bound these quantities in any concise manner.

REFERENCES

Mitchell, Toby J. and Turnbull, Bruce W. (1978). "Exploratory Analysis of Disease Prevalence Data From Survival/Sacrifice Experiments," Biometrics, Vol. 34, pp. 555-570.

Mitchell, Toby J. and Turnbull, Bruce W. (1979). "Log-Linear Models in the Analysis of Disease Prevalence Data From Survival/Sacrifice Experiments," Biometrics, Vol. 35, pp. 221-234.

Neyman, J. and Scott, E. L. (1967). "Statistical aspect of the problem of carcinogenesis," Proc. Fifth Berkeley Symposium on Math. Stat. and Prob. (L. Le Cam and J. Neyman, eds.), University of California Press, Berkeley and Los Angeles, Vol. 4, pp. 745-776.

LARGE DEVIATIONS, TESTS, AND ESTIMATES

R. R. Bahadur
S. L. Zabell

Department of Statistics
University of Chicago*
Chicago, Illinois

J. C. Gupta

Indian Statistical Institute
Calcutta, India

SUMMARY

Let $\{x_n: n = 1,2,\ldots\}$ be a sequence of random elements, for each n let A_n be an event which depends only on $(x_1,\ldots,x_n)$, and let $a_n = n^{-1} \log P(A_n)$. A lower bound for $\liminf_{n \to \infty} a_n$ is described. Various lower bounds in the large deviations literature are obtainable from the present one. The present bound is established, in effect, by application of the Neyman-Pearson lemma, and attainment of the bound is shown to be related to the existence of point estimates which are efficient in a large deviations sense. The paper is essentially an exposition and generalization of certain formulations and arguments of Bahadur, Donsker and Varadhan, Gupta, and Sievers.

*Support for this work was provided in part by National Science Foundation Grant NO. MCS76-81435.

ISBN: 0-12-166650-6

I. INTRODUCTION

The problem of large deviations needs no description here; the reader may wish to see Groeneboom, Oosterhoff, and Ruymgaart (1979), Bahadur and Zabell (1979), and references given in these two papers. These papers of Groeneboom _et al_, and Bahadur and Zabell, are henceforth referred to as [GOR] and [BZ] respectively. It is stated in Section 1 of [BZ], in terms of a particular framework, that obtaining adequate asymptotic lower bounds for large deviation probabilities seems much easier than obtaining adequate upper bounds, and that attainment of adequate lower bounds is related to the existence of asymptotically efficient tests and estimates. We argue here that these statements concerning large deviation probabilities, tests, and estimates are in fact valid in general frameworks.

Theorem 2.1 in Section 2 provides an asymptotic lower bound for virtually any large deviation probability. The proof of this theorem is essentially a simple application of the Neyman-Pearson lemma, and it is clear from the proof that the bound is attained only if the probability being bounded is the size of an asymptotically optimal critical region in some testing problem. Several (but not all) general lower bounds in the present literature of large deviations can be obtained from Theorem 2.1. Some applications of Theorem 2.1 are considered in Section 3. It is pointed out that in examples such as Examples 3.1-3.4 attainment of the bound provided by Theorem 2.1 is equivalent to certain sample statistics being asymptotically efficient estimates in a large deviations sense. The connection with formal estimation theory is made explicit in Section 4. Theorem 4.1 of Section 4 provides an asymptotic lower bound for large deviation probabilities for any consistent estimate in virtually any estimation problem. Some cases where the bound of Theorem 4.1 is attained are described, and the relation of Section 4 to some previous work in estimation theory is pointed out.

II. A LOWER BOUND FOR LARGE DEVIATION PROBABILITIES

Let S be a space of points s, $\mathcal{A}$ a σ-field of subsets of S, and P a probability measure on $\mathcal{A}$. For each $n = 1,2,\ldots$ let $\mathcal{B}_n$ be a σ-field such that

$$\mathcal{B}_n \subset \mathcal{A} \qquad (n = 1,2,\ldots) \tag{2.1}$$

For each n let A_n be a subset of S, and let $\alpha_n = P(A_n)$ if A_n is $\mathcal{B}_n$-measurable; if A_n is not $\mathcal{B}_n$-measurable, let

$$\alpha_n = P_n(A_n) \tag{2.2}$$

where P_n is the inner measure on S determined by the restriction of P to $\mathcal{B}_n$, i.e., $P_n(B) = \sup\{P(C)\colon\ C \in \mathcal{B}_n,\ C \subset B\}$ for all $B \subset S$. Theorem 2.1 of this section provides a lower bound for $\liminf_{n \to \infty} n^{-1} \log \alpha_n$.

We may think of s as a sequence $(x_1, x_2, \ldots \text{ ad inf})$ of random elements taking values in a space X, and of $\{A_n\}$ as a sequence of events such that A_n depends on s only through $(x_1,\ldots,x_n)$. In this interpretation, the x_n are not necessarily i.i.d., and (cf., e.g., [GOR], Remark 3.1) the events A_n are not necessarily measurable. With S the set X^∞ of all possible sequences s, and $\mathcal{A}$ a given σ-field of sets of S, we may think of $\mathcal{B}_n$ as the subfield of $\mathcal{A}$

induced by the mapping which takes $s = (x_1, x_2, \ldots$ ad inf) into $(x_1, \ldots, x_n)$. In this case, $\mathcal{B}_n \subset \mathcal{B}_{n+1}$ for all n, but this monotonicity property is not required here or in Section 4. Indeed, there are examples where A_n depends on $(x_1, \ldots, x_n)$ only through a statistic $U_n(x_1, \ldots, x_n)$, U_n is not a function of U_{n+1}, and it is advantageous, in Theorems 2.1 and 4.1, to take $\mathcal{B}_n$ to be the subfield of $\mathcal{A}$ induced by the mapping which takes s into U_n. It might be added here that such statistics U_n are not necessarily real valued; see Example 3.5.

The method we use to bound α_n may be outlined as follows. Introduce a probability Q on $\mathcal{A}$. For each n, regard A_n as a critical region for testing the hypothesis that P obtains against the alternative that Q obtains. Let B_n be a Neyman-Pearson critical region for testing P against Q when the sample space is $(S, \mathcal{B}_n)$ such that the power of B_n is not greater than the power of A_n, i.e.,

$$Q_n(A_n) \geq Q_n(B_n). \tag{2.3}$$

Then $\alpha_n = P_n(A_n) \geq P_n(B_n)$. Since the set B_n has a special structure, it is possible to estimate $n^{-1} \log P_n(B_n)$ accurately for large values of n. An asymptotic bound for $n^{-1} \log \alpha_n$ is thus obtained, the bound being dependent on the more or less arbitrary Q. By varying Q suitably, the best bound available by the method is obtained. We note that this last lower bound can be attained essentially only if, for some suitable Q, A_n is asymptotically equivalent to B_n for testing P against Q. We note also that, because A_n is arbitrary, the method cannot be used to bound α_n from above, even if Q and B_n are chosen so that (2.3) is reversed.

The method described above is used explicitly in Bahadur (1960, 1967, 1971) in the context of asymptotically efficient estimation. The method is used also in certain purely probabilistic contexts implicitly by Donsker and Varadhan (1975, 1976), and explicitly by Sievers (1976). In these papers of Bahadur, Donsker and Varadhan, and Sievers, each Q deployed evidently satisfies, or is shown to satisfy, certain conditions which facilitate construction of B_n and estimation of $P(B_n)$. Since we are at present concerned only with lower bounds, it suffices that Q satisfy the following condition:

$$\liminf_{n \to \infty} Q_n(A_n) > 0. \tag{2.4}$$

Suppose first that

$$Q \ll P \text{ on } \mathcal{B}_n \tag{2.5}$$

for all sufficiently large n. For each n such that (2.5) holds let $r_n(s)$ be a $\mathcal{B}_n$-measurable function such that $0 \le r_n(s) < \infty$ and such that

$$Q(ds) = r_n(s)\, P(ds) \quad \text{on } \mathcal{B}_n. \tag{2.6}$$

Then r_n is the likelihood ratio statistic for testing P against Q when the sample space is $(S, \mathcal{B}_n)$. For any constant t, $-\infty \le t \le \infty$, let

$$C_n(t) = \left\{ s:\ n^{-1} \log r_n(s) \le t \right\} \tag{2.7}$$

It follows easily from (2.6) and (2.7) (see Bahadur and Raghavachari (1972), page 133) that $\mathcal{Q}(C_n(t)) \le \exp(nt)$; hence $\mathcal{Q}(C_n(t)) \to 0$ as $n \to \infty$ if $t < 0$. On the other hand, $\mathcal{Q}(C_n(\infty)) = 1$ for all n. It follows hence that

$$\inf\left\{ t: \lim_{n \to \infty} \mathcal{Q}(C_n(t)) = 1\right\} = K(\mathcal{Q},P) \text{ say} \tag{2.8}$$

is well-defined, and that $0 \le K \le \infty$. If (2.5) fails for infinitely many n we define K to be $+\infty$.

We note that K as defined here depends only on the initial framework S, $\mathcal{A}$, $\{\mathcal{B}_n\}$, P and on the measure $\mathcal{Q}$ introduced for present purposes. It can be seen from Sections 6 and 7 of Bahadur and Raghavachari (1972), and from the following Section 3, that often K is a Kullback-Leibler information number and $n^{-1} \log r_n(s) \to K$ in $\mathcal{Q}$-probability. Theorem 2.1 and Corollary 2.1 do not require that the properties just stated hold, and so are applicable in contexts such as that of Section 5 of Bahadur and Raghavachari (1972).

THEOREM 2.1. If $\{A_n\}$ is a sequence of subsets of S and $\mathcal{Q}$ a probability measure on $\mathcal{A}$ such that (2.4) holds then

$$\liminf_{n \to \infty} n^{-1} \log P_n(A_n) \ge -K(\mathcal{Q},P). \tag{2.9}$$

PROOF. Choose and fix a $\mathcal{Q}$ such that (2.4) holds, and abbreviate $K(\mathcal{Q},P)$ to K. Since (2.9) holds trivially if $K = \infty$, suppose that $0 \le K < \infty$. Choose and fix a t such that $t > K$. Let n be restricted to values so large that (2.5) holds, and abbreviate $C_n(t)$ defined by

(2.6) - (2.7) to C_n. Let B_n denote the complement of C_n. It then follows from $t > K$ by (2.7) and (2.8) that

$$\lim_{n \to \infty} Q(B_n) = 0. \tag{2.10}$$

Consider a particular n. Suppose that A_n is a $\mathcal{B}_n$-measurable set. Then

$$\begin{aligned} P(A_n) &\geq P(A_n \cap C_n) \\ &= \int_{A_n \cap C_n} P(ds) \\ &\geq \int_{A_n \cap C_n} r_n \exp(-nt)\, P(ds) \quad \text{by (2.7)} \\ &= e^{-nt}\, Q(A_n \cap C_n) \quad \text{by (2.6)} \\ &\geq e^{-nt}\, [Q(A_n) - Q(B_n)]. \end{aligned} \tag{2.11}$$

Since the $\mathcal{B}_n$-measurable A_n in (2.11) is arbitrary, and B_n is a fixed $\mathcal{B}_n$-measurable set, it follows easily from (2.11) that

$$P_n(A_n) \geq e^{-nt}[Q_n(A_n) - Q(B_n)] \tag{2.12}$$

for any $A_n \subset S$. It follows from (2.4), (2.10) and (2.12) that the left hand side in (2.9) is not less than $-t$. Since $t > K$ is arbitrary, (2.9) holds. This completes the proof.

In the trivial case when $\alpha_n = P_n(A_n)$ is bounded away from 0, (2.9) with $Q = P$ yields the trivial conclusion that $n^{-1} \log \alpha_n \to 0$. In case $\alpha_n \to 0$ very rapidly, (2.9) is non-trivial only if $K < \infty$; then (2.5) holds for all sufficiently large n, and (2.4) and (2.5) become opposing constraints: for large n, $P_n(A_n)$ is very small, so $Q_n(A_n)$ tends to be very small according to (2.5) but not very small according to (2.4). In any case, the best bound available from Theorem 2.1 is obtained by choosing a Q which, subject to (2.4), is as close to P as possible in the sense that $K(Q,P)$ is as small as possible. As is well known, the Kullback-Leibler information number is a sort of squared distance; see, e.g., Bahadur (1971), pages 10-11, and Csiszár (1975).

We observe next that (2.12) holds provided only that (2.5) holds and $B_n = B_n(t)$; consequently, the opposing constraint (2.4) may be varied to obtain various asymptotic lower bounds for α_n. A complete discussion of the issues is given in Gupta (1972) in terms of randomized tests of P against Q in the case when s represents an i.i.d. sequence. Here we consider only the question whether (2.9) continues to hold with (2.4) relaxed to the natural condition

$$\lim_{n \to \infty} n^{-1} \log Q_n(A_n) = 0. \tag{2.13}$$

The following Corollary 2.1 shows that in general the constraint (2.5) must then be replaced by a stronger one.

COROLLARY 2.1. Suppose that $K(Q,P) < \infty$. Then (2.9) holds for every sequence $\{A_n\}$ satisfying (2.13) if and only if the following condition is satisfied: For each $t > K(Q,P)$,

$$\limsup_{n \to \infty} n^{-1} \log Q(B_n(t)) < 0 \tag{2.14}$$

where $B_n(t)$ is the complement of $C_n(t)$ defined by (2.7).

PROOF. Suppose first that $\{A_n\}$ satisfies (2.13), and (2.14) holds for each $t > K$. Choose such a t. It follows from (2.13) and (2.14) that the ratio $Q(B_n)/Q_n(A_n)$ is well-defined for all sufficiently large n and that this ratio $\to 0$ as $n \to \infty$. It follows hence from (2.12) and (2.13) that the left hand side in (2.9) is not less than $-t$; hence (2.9) holds. Suppose next that there exists a $t > K$ such that (2.14) fails. Then there exists an increasing sequence $m_1, m_2, \ldots$ of positive integers such that $B_n = B_n(t)$ is well-defined for $n \geq m_1$, and

$$\lim_{j \to \infty} m_j^{-1} \log Q(B_{m_j}) = 0. \tag{2.15}$$

We observe next that, for each sufficiently large n,

$$\begin{aligned} P(B_n) &= \int_{B_n} P(ds) \\ &\leq \int_{B_n} r_n \exp(-nt)\, P(ds) \quad \text{by (2.7)} \\ &= e^{-nt}\, Q(B_n) \quad \text{by (2.6)} \\ &\leq e^{-nt}. \end{aligned} \tag{2.16}$$

For each $n = 1,2,\ldots$ let $A_n = B_n$ if $n = m_j$ for some j, and let $A_n = S$ otherwise. It then follows from (2.15) that (2.13) is satisfied by $\{A_n\}$, and from (2.16) and $t > K$ that (2.9) does not hold. This completes the proof.

III. SOME EXAMPLES

In each of the examples of this section much more is known about large deviation probabilities than is presented here. Our object is considering these examples is to illustrate the nature, scope, and limitations of Theorem 2.1.

Let X be a space of points x, and $\mathcal{C}$ a σ-field of subsets of X. Let p and q be probability measures on $\mathcal{C}$. In case $q << p$ on $\mathcal{C}$, let $\kappa(q,p) = \int_X [\log r(x)] q(dx)$, where r is a $\mathcal{C}$-measurable version of dq/dp; if q is not dominated by p, let $\kappa(q,p) = +\infty$. In this section and the following one we shall sometimes refer to $\kappa(q,p)$ as just defined as "the average information for testing p against q." In such phrases, the space X and the field $\mathcal{C}$ will be clear from the context. κ is, of course, a Kullback-Leibler number.

In each example of this section, $(S,\mathcal{A})$ is the product of countably many copies of a measurable space $(X,\mathcal{C})$ so that $s = (x_1,x_2,\ldots \text{ ad inf})$ is a sequence of elements of X. Unless otherwise stated, $\mathcal{B}_n$ is the subfield of $\mathcal{A}$ induced by the mapping $s \to (x_1,\ldots,x_n)$. The reader may verify that if $P = p^\infty$ and $\mathcal{Q} = q^\infty$, so that $\{x_n\}$ is an i.i.d. process under P and also under $\mathcal{Q}$, then $K(\mathcal{Q},P)$ of Section 2 always equals $\kappa(q,p)$ as defined above.

<u>Example 3.1</u>. Suppose that X is a finite set, say $X = \{1,2,\ldots,k\}$, with $k \geq 2$, and $\mathcal{C}$ is the class of all subsets of X. Let $p_1,p_2,\ldots,p_k$ be given positive constants with $\sum_{j=1}^{k} p_j = 1$. Let s

represent an i.i.d. sequence $\{x_n\}$ with each x_n taking the value j with probability p_j, $j = 1,\ldots,k$. Let P be the resulting probability measure on S.

For each s and n, let $f_j^{(n)}(s)$ be the number of x_i with $1 \le i \le n$ such that $x_i = j$, and let

$$T_n(s) = n^{-1}\left(f_1^{(n)}(s),\ldots,f_k^{(n)}(s)\right). \tag{3.1}$$

Let V_1 denote the set of all points $v = (v_1,\ldots,v_k)$ in R^k such that $v_j \ge 0$ for all j and $\sum_{j=1}^{k} v_j = 1$, and let V_1 be equipped with the usual topology. We may think of V_1 as the set of all probability measures on X, $p = (p_1,\ldots,p_k)$ as the actual probability measure, and T_n as the empirical probability measure based on $(x_1,\ldots,x_n)$. Let U be a non-empty open subset of V_1, and let

$$a_n(U) = n^{-1}\log P(T_n(s) \in U). \tag{3.2}$$

Choose a v in U, and let Q be the distribution on S when the i.i.d. x_i are distributed according to v. For each n, let

$$A_n = \left\{s\colon\ T_n(s) \in U\right\}. \tag{3.3}$$

Since U is open, it follows from (3.3) and the law of large numbers that

$$\lim_{n\to\infty} Q(A_n) = 1. \tag{3.4}$$

In the present case the average information for testing p against v is

$$I_1(v) = \sum_{j=1}^{k} v_j \log(v_j/p_j) \tag{3.5}$$

with $0 \log 0 = 0$. It follows hence from (3.2) - (3.4) by Theorem 2.1 that $\liminf_{n \to \infty} a_n(U) \geq - I_1(v)$. Since $v \in U$ is arbitrary,

$$\liminf_{n \to \infty} a_n(U) \geq - \inf \left\{ I_1(v): \quad v \in U \right\} \tag{3.6}$$

$$= L_1(U) \quad \text{say.}$$

It is known (see, e.g., Example 5.4 in Bahadur (1971) or Example 6.1 in [BZ]) that in fact the bound (3.6) is exact, i.e.,

$$\lim_{n \to \infty} a_n(U) = L_1(U) \tag{3.7}$$

for every open U.

Now for each n choose and fix a function Z_n of $(x_1,\ldots,x_n)$ taking values in V_1 so that the sequence $\{Z_n\}$ is a consistent estimate of v, i.e., if some v in V_1 obtains then $Z_n(s) \to v$ in probability. Let $b_n(U)$ be defined by (3.2) with T_n replaced by Z_n. Then, by replacing T_n with Z_n in the argument leading to (3.6) we have $\liminf_{n \to \infty} b_n(U) \geq L_1(U)$ for every open U. It follows hence from (3.7) that if the distribution over X is entirely unknown then $\{T_n\}$ is an asymptotically optimal estimate, in a strong large deviations sense, in the class of all consistent estimates, at least whenever a distribution p with positive coordinates obtains: for any open U whose closure does not contain p, $P(T_n \in U) \to 0$ at the fastest possible exponential rate. This conclusion is hardly unexpected or inspiring; we state it here to point out that the conclusion depends on the attainment (see (3.6), (3.7)) of the bound provided by Theorem 2.1. We might add that in more complicated frameworks it is difficult to prove that a plausible estimate attains the relevant lower bound; indeed, there are examples where the plausible estimate attains the relevant lower bound only for certain special open sets (see Examples 3.3 and 3.4 below).

In the present finite multinomial case, conclusions concerning T_n much deeper than (3.7) were obtained by Hoeffding (1965), and applied to the comparison of tests of statistical hypotheses. Further statistical applications of Hoeffding's large deviation estimates are given in Gupta (1972). A continuation of Hoeffding's probabilistic work on the multinomial is presented by Reeds in this Volume.

Example 3.2. Let $X = \{1,2,\ldots,k\}$, C, and V_1 be the same as in Example 3.1, but suppose now that $(x_1,x_2,\ldots)$ is a Markov chain. To be specific, let $\{p_{ij}\}$ be a $k \times k$ matrix with each row in V_1, let $u = (u_1,\ldots,u_k)$ be a point in V_1, and let P be determined by the conditions $P(x_1 = i) = u_i$ and $P(x_{n+1} = j | x_1,x_2,\ldots x_{n-1},x_n = i) = p_{ij}$ for all $i, j \in X$ and all $n = 1,2,\ldots$. It is assumed that each element of the transition probability matrix $\{p_{ij}\}$ is positive.

For each s and n let $f_j^{(n)}(s)$ for $j = 1,2,\ldots,k$ be defined as in Example 3.1. Let U be an open subset of V_1 and let $a_n(U)$ be defined by (3.1) and (3.2). In order to apply Theorem 2.1 to the present $a_n(U)$ it is natural to introduce measures Q on S such that $(x_1, x_2, \ldots)$ is a Markov chain under Q. To proceed exactly as in Example 3.1, we choose a point $v = (v_1, \ldots, v_k)$ in U, and require of Q that (i) $T_n \to v$ in Q-probability. Then (3.4) will hold for A_n defined by (3.3). In the present case, it is convenient to let $\mathcal{B}_n$ be the subfield of $\mathcal{A}$ induced by the mapping $s \to (x_1, \ldots, x_{n+1})$. Let us assume that (ii) $Q(x_1 = i) = u_i$ for $i = 1, 2, \ldots, k$, i.e., that Q and P have the same initial distribution, for otherwise (2.5) might fail for every n and thus $K(Q, P)$ become $+\infty$. With (ii) in force, Q is determined by its one-step transition probability matrix, say $\{q_{ij}\}$. Next, for each n let $g_{ij}^{(n)}(s)$ be the number of indices m with $1 \le m \le n$ such that $x_m = i$ and $x_{m+1} = j$, $(i, j = 1, \ldots, k)$. Since $p_{ij} > 0$ for all i and j, it follows easily from (ii) and the present definition of $\mathcal{B}_n$ that (2.5) holds, and that with r_n defined by (2.6) we have

$$n^{-1} \log r_n(s) = \sum_{i,j=1} \left[g_{ij}^{(n)}(s)/n \right] \log \left(q_{ij}/p_{ij} \right) \tag{3.8}$$

with Q-probability 1. Let us assume that (iii) for each i and j, $g_{ij}^{(n)}(s)/n \to v_i q_{ij}$ in Q-probability. It will then follow from (3.8) that $K(Q, P)$ equals

$$\sum_{i,j=1}^{k} \left(v_i q_{ij} \right) \log \left(q_{ij}/p_{ij} \right) = J(v, \{q_{ij}\}) \quad \text{say.} \tag{3.9}$$

Since $\sum_{j=1}^{k} g_{ij}^{(n)}(s) \equiv f_i^{(n)}(s)$, and $\sum_{j=1}^{k} v_i q_{ij} \equiv v_i$, it is plain that (iii) implies that condition (i) above is automatically satisfied. We conclude that if (iii) is satisfied then, with J given by (3.9),

$$\liminf_{n \to \infty} a_n(U) \geq -J(v, \{q_{ij}\}). \tag{3.10}$$

It is clear that, although J is well-defined for all v and $\{q_{ij}\}$, conditions (i) and (iii) require that v be a stationary distribution for $\{q_{ij}\}$, i.e.,

$$\sum_{i=1}^{k} v_i q_{ij} = v_j \qquad (j = 1,2,\ldots,k). \tag{3.11}$$

Let M(v) be the set of all $\{q_{ij}\}$ which satisfy (3.11), and let

$$I_2(v) = \inf \left\{ J(v,\{q_{ij}\}) \; : \; \{q_{ij}\} \in M(v) \right\}. \tag{3.12}$$

We now show that there always exists a $\{q^*_{ij}\}$ in M(v) such that

$$I_2(v) = J(v,\{q^*_{ij}\}) \tag{3.13}$$

and such that condition (iii) above is satisfied when $\{q^*_{ij}\}$ is the transition probability matrix of $\mathcal{Q}$. Suppose first that all coordinates of v are positive. Let X^2 be the set of all pairs (i,j) with $i,j \in X$. Then $\{q_{ij}\} \to \{v_i q_{ij}\}$ is a one-to-one correspondence between M(v) and the set of all probability distributions $\{\beta_{ij}\}$ on X^2 such that $\sum_j \beta_{ij} = v_i$ and $\sum_i \beta_{ij} = v_j$ for all i and j. We observe next that J defined by (3.9) equals the average information for testing $\{v_i p_{ij}\}$ against $\{v_i q_{ij}\}$. Since $v_i p_{ij} > 0$ on X^2, it follows (see, e.g., Ireland and Kullback (1968), and Csiszár (1975)) that J is minimized by a unique $\{q^*_{ij}\}$ in M(v), and that there exists positive constants ρ_i and δ_j such that $q^*_{ij} = \rho_i \cdot p_{ij} \cdot \delta_j$ on X^2. Since each element of $\{q^*_{ij}\}$ is positive, and v is its stationary distribution, condition (iii) does hold if Q is governed by $\{q^*_{ij}\}$. Suppose now that exactly m of the coordinates of v are positive, with $1 \le m < k$, say $v = (v_1, \ldots, v_m, 0, \ldots, 0)$. In this case (3.11) implies that (iv) $q_{ij} = 0$ for $1 \le i \le m$ and $j > m$. Let $r_{ij} = p_{ij}/(\sum_{j=1}^{m} p_{ij})$ for $i,j = 1,\ldots,m$. It then follows from condition (iv) that, for all $\{q_{ij}\}$ in $M(v)$,

$$J(v,\{q_{ij}\}) = \sum_{i,j=1}^{m} \left(v_i q_{ij}\right) \log \left(q_{ij}/r_{ij}\right) + H(v). \qquad (3.14)$$

The variable term on the right-hand side of (3.14) is of exactly the same form as J, with X replaced by $X_0 = \{1,\ldots,m\}$, $\{p_{ij}\}$ replaced by $\{r_{ij}\}$, and v replaced by $v_0 = (v_1,\ldots,v_m)$. Since each coordinate of v_0 is positive, it follows as above that, in the class of all transition matrices for the state space X_0 which have v_0 as a stationary distribution, there exists a $\{q^*_{ij}\}$ such that $q^*_{ij} > 0$ for $1 \le i, j \le m$ which minimizes the right-hand side of (3.14). To complete the definition of $\{q^*_{ij}\}$ for the original state space, let $q^*_{ij} = 0$ for $1 \le i \le m$ and $j > m$, in accordance with condition (iv); and for $m + 1 \le i \le k$ let $q^*_{ij} = 1$ if $j = 1$ and 0 if $j \ne 1$. The effect of this last specification is that, when Q is governed by $\{q^*_{ij}\}$, $\{m + 1,\ldots,k\}$ is the set of all transient states; consequently, whatever the initial distribution, condition (iii) is satisfied.

By putting $\{q_{ij}\} = \{q^*_{ij}\}$ in (3.10) it follows that, with I_2 defined by (3.12), the left-hand side of (3.10) is not less than $-I_2(v)$. Since $v \in U$ is arbitrary we conclude that

$$\liminf_{n \to \infty} a_n(U) \geq -\inf\left\{I_2(v) : v \in U\right\} \tag{3.15}$$

$$= L_2(U) \quad \text{say}.$$

The above is an exposition, in terms of a finite state space X, of certain arguments of Donsker and Varadhan (1975, 1976) who obtained (3.15) first for the case when X is compact metric and then for any Polish state space. The case when X is infinite presents formidable technical difficulties. The lower bound of Donsker and Varadhan is therefore a remarkable achievement, especially since they show, with X a Polish space, that here $-I_2$ is the correct entropy function in the following sense: If C is a nonempty closed subset of V_1, and $a_n(C)$ and $L_2(C)$ are defined as above with U replaced by C, then $\limsup_{n \to \infty} a_n(C) \leq L_2(C)$. Donsker and Varadhan obtain this last upper bound by using minimax theorems. Minimax theorems are also used in [GOR], and implicitly in [BZ], to obtain upper bounds for certain large deviation probabilities.

It can be shown (e.g., by using the methods and results of Boza (1971)) that, in the present finite case, the bound (3.15) is exact, i.e.,

$$\lim_{n \to \infty} a_n(U) = L_2(U) \tag{3.16}$$

for every nonempty open $U \subset V_1$. It follows from (3.16), as in Example 3.1, that the present T_n is an asymptotically optimal estimate of the stationary distribution of a chain with virtually unknown transition probabilities.

In the following Examples 3.3 and 3.4, $(X,\mathcal{C})$ is a general measurable space, $S = X^{\infty}$, $\mathcal{A} = \mathcal{C}^{\infty}$, and $P = p^{\infty}$, where p is a given probability measure on $\mathcal{C}$. We refer the reader to [GOR] and [BZ] for the backgrounds of, and interrelationships between, the large deviation problems considered in Examples 3.3 and 3.4. Examples 3.3 and 3.4 are generalizations, in directions different from that of Example 3.2, of the basic Example 3.1.

Example 3.3. Let V_1 be the set of all probability measures on $\mathcal{C}$, and let V_1 be equipped with a topology τ_0. For each s and n let M_n denote the empirical probability measure based on $(x_1,\ldots,x_n)$, i.e., for any $A \subset X$, $M_n(s,A) =$ (number of indices i with $1 \le i \le n$ and $x_i \in A)/n$. Then, for each s and n, M_n is an element in V_1. Let us say that the WLLN holds if $q \in U \subset V_1$ and U is open, then with $Q = q^{\infty}$, $Q_n(M_n \in U) \to 1$ as $n \to \infty$. Suppose that the WLLN does hold; this is a condition on $(X,\mathcal{C})$ and on the given topology τ_0 on V_1. It then follows from Theorem 2.1, exactly as in Example 3.1, that

$$\liminf_{n \to \infty} n^{-1} \log P_n(M_n \in U) \ \ge\ \sup\{-\kappa(q,p) : q \in U\} \tag{3.17}$$

for every nonempty open set $U \subset V_1$. This inequality is stated in Section 7 of [BZ]. An even more general inequality was obtained previously, by exactly the same method as the present one, by Sievers (1976).

Now consider the particular topology τ which is introduced in [GOR] and shown there to be most useful, in several different ways, in the present context. Let π denote a finite $\mathcal{C}$-measurable partition of X, i.e., $\pi = \{C_1,\ldots,C_k\}$ where $C_i \in \mathcal{C}$, $C_i \cap C_j = \phi$ for

$i \neq j$ $(i,j = 1,\ldots,k)$, and $\bigcup_1^k C_i = X$. For $m,q \in V_1$ let $d_\pi(m,q) = \max\{|m(C_i) - q(C_i)| : 1 \leq i \leq k\}$, and let $B_\pi(q,\delta) = \{m : m \in V_1, d_\pi(m,q) < \delta\}$ for $\delta > 0$. Then τ is the smallest topology in which $B_\pi(q,\delta)$ is open for every π, q, and δ. This is a very large Hausdorff topology, but, as is pointed out in the following paragraph, the WLLN and even a parallel SLLN hold in τ. It follows, in particular, that (3.17) holds for all τ-open sets U. This conclusion is established in [GOR] by a different method under the assumption that X itself is a Hausdorff space and $\mathcal{C}$ is the Borel field.

Choose q in V_1, and let U be a τ-open set containing q. As noted in [GOR], sets of the form B are a basis for τ. Hence there exist π, m, and $\varepsilon > 0$ such that $q \in B_\pi(m,\varepsilon) \subset U$. Since d_π is a pseudometric, it follows that there exists $\delta > 0$ such that $B_\pi(q,\delta) \subset B_\pi(m,\varepsilon)$. Hence $B_\pi(q,\delta) \subset U$. It now follows from the WLLN and the SLLN for the finite multinomial, as desired, that with $\mathcal{Q} = q^\infty$, $\mathcal{Q}_n\{s : M_n \in U\} \to 1$ as $n \to \infty$, and that there exists an $\mathcal{A}$-measurable set of sequences s, of $\mathcal{Q}$-measure 1, such that for each s in the set, $M_n \in U$ for all sufficiently large n.

Two special cases of interest are the following. Case 1: X is a Polish space, $\mathcal{C}$ is the Borel field, and V_1 is equipped with the weak topology. Case 2: X is the k-dimensional Euclidean space, $\mathcal{C}$ is the Borel field, and the topology on V_1 is that of uniform convergence of probability distribution functions. As is shown in [GOR], the topologies in either case are smaller than τ, so (3.17) holds for all open sets in each case. We note that since the SLLN holds in τ, it holds in Cases 1 and 2 also. Since V_1 is metrizable in these cases, it follows that the strong consistency theorems of Varadarajan (Case 1), and of Glivenko and Cantelli (Case 2), are valid. We think that this proof of the Varadarajan-Glivenko-Cantelli theorems is yet another demonstration of the utility of the τ topology of [GOR].

The connection with asymptotically efficient estimation (see Example 3.1) suggests that (3.17) might in fact be exact for every open U, at least in Case 1 above, with X a finite dimensional Euclidean space. Example 7.1 in [BZ] shows, however, that this suggestion is false even if X is the real line and p represents the uniform distribution on the interval $(0,1)$. It should be added here that the failure just described is, perhaps, not a failure of M_n as an estimator but of Theorem 2.1 itself.

Example 3.4. Suppose that X is a locally convex (Hausdorff) topological vector space, and $\mathcal{C}$ is the Borel field. Let q be a probability measure on $\mathcal{C}$. We shall say that q has mean $m(q)$ if $m(q)$ is the (necessarily unique) point in X such that, for every continuous linear functional f on X, the integral $\int_X f(x)\, q(dx)$ exists and equals $f(m(q))$.

For each $s = (x_1, x_2, \ldots)$ let $T_n(s) = (x_1 + \ldots + x_n)/n$. Let us say that the WLLN holds if, for each q such that $m(q)$ exists, with $Q = q^\infty$ we have $\lim_{n \to \infty} Q_n(T_n(s) \in U) = 1$ for every open U containing $m(q)$. For any set $U \subset X$ let $\alpha_n(U) = P_n(T_n(s) \in U)$, and $a_n(U) = n^{-1} \log \alpha_n(U)$. Assume that the WLLN holds. It then follows from Theorem 2.1 that

$$\liminf_{n \to \infty} a_n(U) \geq -\inf\{K(q,p) : m(q) \in U\} \tag{3.18}$$

for every nonempty open $U \subset X$.

The bound (3.18) is closely related to, and often identical with, an asymptotic lower bound for $a_n(U)$ obtained in [BZ]. To be specific, if Assumptions 1, 2, and 3 of [BZ] (which do not include the WLLN) hold, it follows from Theorem 3.3 of [BZ] that (3.18) is the inequality $\liminf_{n \to \infty} a_n(U) \geq \mathrm{lan}(U)$ of [BZ], where $\mathrm{lan}(A)$ denotes the Lanford entropy of the set A.

Suppose now that X is a Polish space. Then Assumptions 1, 2, and 3 of [BZ] are satisfied, and by Theorem 2.3 of [BZ] the inequality (3.18) is exact whenever U is an open convex set. This conclusion, together with Theorem 2.1, implies that $T_n(s)$ is an asymptotically optimal estimate at least in the following sense: If it is known only that x is distributed in X according to some p such that $m(p)$ exists, and $Z_n(s)$ is any consistent estimate of m, then, for any open half-space H in X whose closure does not contain $m(p)$, the exponential rate of convergence to zero of $P(Z_n(s) \in H)$ cannot be faster than that of $P(T_n(s) \in H)$. This connection between the conclusions of [BZ] and statistical inference seems especially relevant to the main argument of this paper. It is known that sample means, and half-spaces, are important in testing and estimation problems in exponential families of measures, and [BZ] makes extensive explicit or implicit use of computations characteristic of such families, e.g., finding the Fenchel transform of a cumulative generating function.

It is known that (3.18) is not exact for every open U, e.g., if X is the space of continuous functions on the interval $[0,1]$ of the real line, X is equipped with the topology of uniform convergence, and p is the standard Wiener measure; see Borovkov and Mogul'skii (1978), page 701. Cf., however, Example 4.3 in Section 4.

Suppose next that X is the k-dimensional Euclidean space. It can be shown (e.g., by using certain conclusions of [BZ]) that if $k = 1$, (3.18) is exact for every open U. An important theorem of Bártfai (1978) states in part that this last is the case for any k provided that the moment generating function of p is finite in a neighborhood of the origin.

Example 3.5. Suppose that $x_1, x_2, \ldots$ is an i.i.d. sequence of real valued $N(0,1)$ variables. Let β be a constant, $0 < \beta < 1/2$, and let $T_n(s)$ be the β-trimmed mean of $x_1, \ldots, x_n$. Choose an $\varepsilon > 0$, and let $\alpha_n = P(T_n(s) \geq \varepsilon)$, $a_n = n^{-1} \log \alpha_n$. It is shown in [GOR] that $\lim_{n \to \infty} a_n$ exists, say $-b$, and an explicit general formula for b is obtained. We do not think that Theorem 2.1 by itself is very useful here, even if we let $(S, \mathcal{B}_n)$ represent the sample space of the order statistics which are actually used to compute the present T_n.

Example 3.6. Suppose that $x_1, x_2, \ldots$ is a sequence of real valued random variables such that, for each n, the vector $(x_1, \ldots, x_n)$ has the normal distribution defined by $E_P(x_i) = 0$, $E_P(x_i^2) = 1$, $E_P(x_i x_j) = \rho^{j-i}$ for $1 \leq i \leq j \leq n$, where $-1 < \rho < 1$. Let $T_n = (x_1 + \ldots + x_n)/n$, and for $\varepsilon > 0$ let α_n and a_n be defined as in Example 3.5. It is easy to see that $n^{1/2}\, T_n$ is exactly normally distributed with mean 0 and variance σ_n^2, where $\sigma_n^2 \to \sigma^2 = (1 + \rho)/(1 - \rho)$ as $n \to \infty$; it follows hence that $\lim_{n \to \infty} a_n = -\varepsilon^2/2\sigma^2$. We include this non-problematic example here because it is unusual to see processes such as the present $\{x_n\}$ in the large deviations context, but Theorem 2.1 can be used to show that $\liminf_{n \to \infty} a_n \geq -\varepsilon^2/2\sigma^2$.

IV. LARGE DEVIATIONS AND ASYMPTOTICALLY EFFICIENT ESTIMATION

In this section we consider a space S of points s, σ-field $\mathcal{A}$ of subsets of S, and a set $\{P_\theta : \theta \in \Theta\}$ of probability measures on $\mathcal{A}$. For each $n = 1,2,\ldots$ let $\mathcal{B}_n$ be a σ-field such that $\mathcal{B}_n \subset \mathcal{A}$. Then $\{(S, \mathcal{B}_n, P_\theta) : \theta \in \Theta\}$ for $n = 1,2,\ldots$ is a sequence of statistical experiments with a common parameter space Θ. The index n may be thought of as the cost of observing a sample point in the space $(S, \mathcal{B}_n)$.

Let Γ be a topological space of points γ, and let g be a function on Θ into Γ. The problem considered here is the estimation of $g(\theta)$. For each n let T_n be a $\mathcal{B}_n$-Borel-measurable function on S into Γ. T_n is, of course, to be thought of as a point estimate of g based on the sample space $(S, \mathcal{B}_n)$. The sequence $\{T_n\}$ is said to be a consistent estimate of g if, for each θ in Θ, $P_\theta(T_n(s) \in A) \to 1$ as $n \to \infty$ for each open $A \subset \Gamma$ with $g(\theta) \in A$.

Assume that Γ is metrizable, and choose and fix a metric D which generates the given topology on Γ. For each n, $\varepsilon > 0$, and θ let

$$\alpha_n(\varepsilon,\theta) = P_\theta(D(T_n(s),g(\theta)) \geq \varepsilon). \tag{4.1}$$

It is shown below that, whatever the consistent sequence $\{T_n\}$, α_n cannot tend to zero at rate faster than a certain exponential rate.

Let θ and δ be points in Θ. We write $\delta \ll \theta$ if and only if there exists an integer $m = m(\theta,\delta)$ such that (2.5) holds with $Q = P_\delta$ and $P = P_\theta$ for all $n > m$; in this case, let $K(P_\delta,P_\theta)$ be defined by (2.6) - (2.8). For each θ in Θ and $\varepsilon > 0$, let $\Delta(\varepsilon,\theta) \subset \Theta$ be defined by

$$\Delta(\varepsilon,\theta) = \{\delta : \delta \in \Theta,\ \delta \ll \theta,\ D(g(\delta), g(\theta)) > \varepsilon\}. \tag{4.2}$$

Let

$$b(\varepsilon,\theta) = \inf\{K(P_\delta,P_\theta) : \delta \in \Delta(\varepsilon,\theta)\} \tag{4.3}$$

if $\Delta(\varepsilon,\theta)$ is nonempty and let $b(\varepsilon,\theta) = +\infty$ otherwise. We note that the function b is defined entirely in terms of the framework S, A, $\{\mathcal{F}_n\}$, $\{P_\theta : \theta \in \Theta\}$, $g : \Theta \to \Gamma$, and D.

THEOREM 4.1. If $\{T_n\}$ is a consistent estimate of g and $\{\alpha_n\}$ is defined by (4.1) then

$$\liminf_{n\to\infty} \left\{n^{-1} \log \alpha_n(\varepsilon,\theta)\right\} \geq -b(\varepsilon,\theta) \tag{4.4}$$

for each θ in Θ and $\varepsilon > 0$.

PROOF. Choose and fix θ and ε. We may assume that $b(\varepsilon,\theta) < \infty$. Then $\Delta(\varepsilon,\theta)$ is a nonempty set; choose a δ in this set. Suppose δ obtains. Then, by consistency, and the continuity of the function $D(\gamma,g(\theta))$ on Γ, $D(T_n(s),g(\theta)) \to D(g(\delta),g(\theta))$ in probability. It is plain from (4.2) that $D(g(\delta),g(\theta))$ exceeds ε; hence, with $A_n = \{s: D(T_n(s),g(\theta)) \geq \varepsilon\}$, $P_\delta(A_n) \to 1$ as $n \to \infty$. It now follows from Theorem 2.1 with $P = P_\theta$ and $Q = P_\delta$ by (4.1) that the left-hand side of (4.4) is not less than $-K(P_\delta,P_\theta)$. Since δ is arbitrary, it follows from (4.2) and (4.3) that (4.4) holds. This completes the proof.

It is clear that $b(\varepsilon,\theta) \geq 0$ and that, in regular cases, $b(\varepsilon,\theta) \to 0$ as $\varepsilon \downarrow 0$. In such cases, we usually have

$$b(\varepsilon,\theta) = \frac{1}{2} c(\theta) \cdot \varepsilon^2 + O(\varepsilon^3) \quad \text{as} \quad \varepsilon \downarrow 0, \tag{4.5}$$

where $0 < c(\theta) < \infty$. There are, however, cases where b is of the order ε as $\varepsilon \downarrow 0$; see Bahadur (1960), pages 250-251, and Example 4.1 below. We do not know whether b can be of the exact order ε^4 (say) as $\varepsilon \downarrow 0$.

As an immediate consequence of Theorem 4.1 we have

COROLLARY 4.1. If $\{T_n\}$ is a consistent estimate of g then

$$\liminf_{\varepsilon \to 0} \liminf_{n \to \infty} \left\{(n\varepsilon^2)^{-1} \log \alpha_n(\varepsilon,\theta)\right\} \geq -\frac{1}{2} c(\theta) \tag{4.6}$$

for each θ in Θ for which (4.5) holds.

Theorem 4.1 and Corollary 4.1 are generalizations of certain formulations of Bahadur (1960); see also Bahadur (1967, 1971), Wijsman (1971), and Perng (1978).

The existence and construction of estimates for which (4.4) or at least (4.6) is attained seems to be a difficult problem; the source of the difficulty is, of course, that in general it is difficult to obtain adequate upper bounds for large deviation probabilities (see Sections 2 and 3). It is clear from the proofs, however, that maximum likelihood estimates, and closely related estimates, are plausible candidates for asymptotic optimality at least in the following sense: $\alpha_n(\varepsilon,\theta) = \exp\{-\frac{n}{2}\varepsilon^2[c(\theta) + d_n(\varepsilon,\theta)]\}$, where $\limsup_{n \to \infty} |d_n(\varepsilon,\theta)|$ tends to 0 as

$\varepsilon \downarrow 0$. Asymptotic optimality of the maximum likelihood estimate in this sense is established, in the i.i.d. case, under various assumptions, in Bahadur (1960, 1967); see also Fu (1973, 1975), Perng (1978), Examples 3.1 and 3.2, and Bahadur (1980).

We now describe three examples, of rather unorthodox sorts, in which the bound of Theorem 4.1 is attained for each θ and ε. In each example, s is a sequence $x_1, x_2, \ldots$ of i.i.d. random elements in a space X. Θ is one-dimensional in Example 4.1, infinite dimensional in Example 4.2, and possibly infinite dimensional in Example 4.3. In Examples 4.1 and 4.2, g is real valued, and we let $\Gamma = R^1$ and $D(\gamma_1, \gamma_2) = |\gamma_1 - \gamma_2|$ in these two examples.

Example 4.1. Suppose that $\Theta = (0, \infty)$, that the x_i are uniformly distributed over $(0, \theta)$ when θ obtains, and that $g(\theta) \equiv \theta$. Here

$$b(\varepsilon,\theta) = \begin{cases} -\log[1 - (\varepsilon/\theta)] & \text{if } 0 < \varepsilon < \theta \\ +\infty & \text{if } \varepsilon \geq \theta, \end{cases} \tag{4.7}$$

and with $T_n^* = \max\{x_1, \ldots, x_n\}$ we have

$$n^{-1} \log \alpha_n^*(\varepsilon,\theta) = -b(\varepsilon,\theta) \tag{4.8}$$

for all n, ε, and θ. The verification is omitted.

Example 4.2. Let θ denote a probability distribution function on the real line such that the derivative $\theta'(x) = d\theta(x)/dx$ exists for each x, and θ' is positive and continuous for $-\infty < x < \infty$. Let Θ be the set of all such θ, and for each θ in Θ let $g(\theta)$ be the median of θ, i.e., $g(\theta)$ is the unique solution of $\int_{-\infty}^{g} \theta'(x)dx = 1/2$.

For each n, let T_n^* be the median value in the sample $(x_1,\ldots,x_n)$. As is well known, $\{T_n^*\}$ is then a consistent estimate of g. We shall show that, for each $\varepsilon > 0$ and θ in Θ,

$$\lim_{n \to \infty} n^{-1} \log \alpha_n^*(\varepsilon,\theta) = -b(\varepsilon,\theta). \tag{4.9}$$

This conclusion is a large-deviations version of the conclusion in Pfanzagl (1976) to the effect that T_n^* is an asymptotically optimal nonparametric estimate in the classical sense of asymptotic variances. It is known (see the references cited in Pfanzagl (1976)) that if each θ is restricted to be symmetric about $g(\theta)$ then T_n^* is no longer optimal in the classical sense. We presume that (4.9) also fails in the symmetric case, but at present we do not know.

To find b in the present case, choose and fix θ in Θ and $\varepsilon > 0$. Since $\delta \ll \theta$ for each δ in Θ, and $D(\gamma_1,\gamma_2) = |\gamma_1 - \gamma_2|$, Δ defined by (4.2) is the union of $\Delta_1 = \{\delta : g(\delta) < g(\theta) - \varepsilon\}$ and $\Delta_2 = \{\delta : g(\delta) > g(\theta) + \varepsilon\}$, both the Δ_i being non-empty sets; hence $b = \min\{b_1,b_2\}$, where $b_i = \inf\{K(P_\delta,P_\theta) : \delta \in \Delta_i\}$ for $i = 1,2$.

The transformation $\theta : R^1 \to (0,1)$ defined by $y = \theta(x)$ establishes a one-to-one strictly monotone correspondence between points $x \in R^1$ and points $y \in (0,1)$. The transformation also establishes a one-to-one correspondence between probability distribution functions $\delta(x)$ in Θ and probability distribution functions $v(y)$ such that $v(0) = 0$, $v(1) = 1$, and such that the derivative $v'(y) = dv(y)/dy$ exists and is positive and continuous on $(0,1)$; let V denote the set of all such distribution functions v. The d.f. θ corresponds, of course, to the uniform d.f. on $(0,1)$. If δ

corresponds to v, then $g(\delta) > g(\theta) + \varepsilon$ if and only if $g(v) > \theta(g(\theta) + \varepsilon)$. Since the Kullback-Leibler numbers are invariant under one-to-one transformations (indeed, sufficient transformations) of the sample point, it follows that

$$b_2 = \inf \left\{ I(v) : v \in V, g(v) > m_2 \right\} \tag{4.10}$$

where $I(v)$ is the average information in y for testing the uniform d.f. on $(0,1)$ against the d.f. v, and $m_2 = \theta(g(\theta) + \varepsilon)$. Since $\theta(g(\theta)) = 1/2$, we have $1/2 < m_2 < 1$.

Consider an arbitrary but fixed m, $0 < m < 1$, $m \neq 1/2$, and consider minimizing $I(v)$ over all v in V such that $g(v) = m$. Choose such a distribution function v. Then $I(v) = \int_0^1 v'(y)[\log v'(y)]dy = \int_0^m v' \log v' dy + \int_m^1 v' \log v' dy$. Since $g(v) = m$, $2v'$ is a probability density over $(0,m)$. It follows hence that the first integral is an affine increasing function of the average information in testing the uniform density over $(0,m)$ against the density $2v'$; the first integral is therefore minimized by letting $v' = (2m)^{-1}$ on $(0,m)$. The same argument shows that the second integral is minimized by letting $v' = (2(1-m))^{-1}$ on $(m,1)$. It follows that $g(v) = m\cdot$ implies that $I(v) \geq I(v_m^*)$, where v_m^* is the d.f. with density $1/(2m)$ on $(0,m)$ and $1/2(1-m)$ on $[m,1)$. Although v_m^* is not in V, there are sequences $\{v^{(j)} : j = 1,2,\ldots\}$ in V such that $g(v^{(j)}) = m$ for all j and $I(v^{(j)}) \to I(v_m^*)$. We conclude that $\inf\{I(v) : v \in V, g(v) = m\} = I(v_m^*) = f(m)$ say, where

$$f(m) = \frac{1}{2} \log [4m(1-m)]^{-1}. \tag{4.11}$$

It is interesting to note that f is the average information in a random variable which takes only two values, say 0 and 1, for testing that the probability of 1 is m against the alternative that it is 1/2.

Since f defined by (4.11) is continuous and strictly increasing for $1/2 < m < 1$, and $1/2 < m_2 < 1$, it follows from the preceding paragraph by (4.10) that $b_2 = f(m_2)$. The value of b_1 is found similarly from the preceding two paragraphs. We conclude that, with

$$m_1(\varepsilon,\theta) = \theta[g(\theta) - \varepsilon], \quad m_2(\varepsilon,\theta) = \theta[g(\theta) + \varepsilon], \tag{4.12}$$

and f defined by (4.11), we have

$$b(\varepsilon,\theta) = \min\left\{f(m_1(\varepsilon,\theta)),\ f(m_2(\varepsilon,\theta))\right\}. \tag{4.13}$$

Now let $y_n = \theta(x_n)$ for each n. Then $y_1, y_2, \ldots$ is a sequence of i.i.d. uniform variables in (0,1) when θ obtains. We have $\alpha_n^*(\varepsilon,\theta) = P_\theta(T_n^* \geq g(\theta) + \varepsilon) + P_\theta(T_n^* \leq g(\theta) - \varepsilon) = P(U_n^* \geq m_2) + P(U_n^* \leq m_1)$, where U_n^* is essentially the median value in $(y_1,\ldots,y_n)$. It is straightforward to show (e.g., by expressing $P(U_n^* \geq m_2)$ and $P(U_n^* \leq m_1)$ as large deviation probabilities in binomial distributions) that $n^{-1} \log \alpha_n^*(\varepsilon,\theta) \to -\min\{f(m_1), f(m_2)\}$. It now follows from (4.13), as desired, that (4.9) holds.

Example 4.3. Suppose that X is a separable Banach space of points x with norm $||x||$, and that p_0 is a centered Gaussian probability measure on the Borel sets of X, i.e., corresponding to each continuous linear functional f on X there exists a non-negative constant $\sigma^2(f)$ such that $f(x)$ is normally distributed with mean 0 and variance $\sigma^2(f)$ when p_0 obtains. Let $\Theta = X$, and for each $\theta \in \Theta$, let p_θ denote p_0 translated to θ, i.e., $x - \theta$ is distributed according to p_0 when θ obtains. Let $\Gamma = \Theta$, $g(\theta) \equiv \theta$, and $D(\theta_1,\theta_2) = ||\theta_1 - \theta_2||$. For each n, let $T_n^*(s) = (x_1+\ldots+x_n)/n$. Then $n^{-1} \log \alpha_n^*(\varepsilon,\theta) \to -b(\varepsilon,\theta)$ as $n \to \infty$ for each ε and θ. The verification is omitted.

In concluding this section we describe a slight elaboration of Theorem 4.1 and Corollary 4.1. For each θ in Θ, choose and fix a metric D_θ which generates the given topology on Γ. Let α_n, Δ, and b be defined by (4.1), (4.2), and (4.3) with D replaced throughout by D_θ. Then Theorem 4.1 and Corollary 4.1 remain valid; indeed the proof of Theorem 4.1 is valid with D replaced throughout by D_θ. This elaboration may be worthwhile for the following reason. In many cases, if θ obtains, there is a metric D_θ which is a natural or interesting local metric in the immediate vicinity of $g(\theta)$; in particular, if γ is the estimated value of $g(\theta)$, and $\gamma \doteq g(\theta)$, then $D_\theta^2(\gamma,g(\theta))$ might represent the loss due to the error of estimation. To consider a simple example, suppose that s is a sequence of independent zero-one variables x_i, that the probability of x_i being 1 is θ, that $\Theta = (0,1)$, and that $g(\theta) \equiv \theta$. In this example, $D_\theta(\gamma_1,\gamma_2) \equiv |\gamma_1 - \gamma_2|$, or $D_\theta(\gamma_1,\gamma_2) \equiv |\gamma_1 - \gamma_2|/[\theta(1-\theta)]^{1/2}$, or $D_\theta(\gamma_1,\gamma_2) \equiv |\gamma_1 - \gamma_2|/\min\{\theta,1-\theta\}$ are possible choices. In this example, as in the general case, the resultant functions b and c of Theorem 4.1 and Corollary 4.1 depend on the entire set $\{D_\theta : \theta \in \Theta\}$ of metrics chosen.

REFERENCES

Bahadur, R. R. (1960). Asymptotic efficiency of tests and estimates. Sankhyā 22, 229-252.

Bahadur, R. R. (1967). Rates of convergence of estimates and test statistics. Ann. Math. Statist. 38, 303-324.

Bahadur, R. R. (1971). Some limit theorems in statistics. SIAM, Philadelphia.

Bahadur, R. R. and Raghavachari, M. (1972). Some asymptotic properties of likelihood ratios on general sample spaces. Proc. Sixth Berkeley Symp. Math. Statist. Prob. I, 129-152.

Bahadur, R. R. and Zabell, S.L. (1979). Large deviations of the sample mean in general vector spaces. Ann. Probability 7, 587-621.

Bahadur, R. R. (1980). Asymptotic optimality of the maximum likelihood estimate in the Markov chain case. (To appear).

Bártfai, P. (1978). Large deviations of the sample mean in Euclidean spaces. Mimeograph series No. 78 -13. Department of Statistics, Purdue University.

Borovkov, A. A. and Mogul'skii, A. A. (1978). Probabilities of large deviations in topological spaces - I. Siberian Math. Journal 19, 697-709.

Boza, L. B. (1971). Asymptotically optimal tests for finite Markov chains. Ann. Math. Statist. 42, 1992-2007.

Csiszár, I. (1975). I-divergence geometry of probability distributions and minimization problems. Ann. Probability 3, 146-158.

Donsker, M. D. and Varadhan, S. R. S. (1975). Asymptotic evaluation of certain Markov process expectations for large time - I. Comm. Pure Appl. Math. 27, 1-47.

Donsker, M. D. and Varadhan, S. R. S. (1976). Asymptotic evaluation of certain Markov process expectations for large time - III. Comm. Pure Appl. Math. 29, 389-461.

Fu, J. C. (1973). On a theorem of Bahadur on the rate of convergence of point estimators. Ann. Statist. 1, 745-749.

Fu, J. C. (1975). The rate of convergence of consistent point estimators. Ann. Statist. 3, 234-240.

Groeneboom, P., Oosterhoff, J., and Ruymgaart, F. H. (1979). Large deviation theorems for empirical probability measures. Ann. Probability 7, 553-586.

Gupta, J. C. (1972). Probabilities of medium and large deviations with statistical applications. Ph.D. Thesis, University of Chicago.

Hoeffding, W. (1965). Asymptotically optimal tests for multinomial distributions. Ann. Math. Statist. 36, 369-408.

Ireland, C. T. and Kullback, S. (1968). Contingency tables with given marginals. Biometrika 55, 179-188.

Perng, S. S. (1978). Rate of convergence of estimators based on sample mean. Ann. Statist. 6, 1048-1056.

Pfanzagl, J. (1976). Investigating the quantile of an unknown distribution. Contributions to Applied Statistics (dedicated to Arthur Linder; edited by W. J. Ziegler), 111-126, Birkhäuser Verlag, Basel.

Sievers, G. L. (1976). Probabilities of large deviations for empirical measures. Ann. Statist. 4, 766-770.

Wijsman, R. A. (1971). Lecture notes on estimation. Department of Mathematics, University of Illinois, Urbana.

APPLICATIONS OF CHARACTERISTIC FUNCTIONS IN SOLVING SOME DISTRIBUTION PROBLEMS

Daniel Dugué

Institut de Statistique des
Universités de Paris
Paris, France

I

The well-known T distribution, universally known as Hotelling's T, can be obtained in a very simple way through characteristic functions and, at the same time, it can be proved that T is independent of a Wilks-Wishart matrix.

Let us consider $X_1,\ldots,X_n,X$ n+1 k-dimensional Gaussian vectors $(n\geq k)$, independent, with zero mean and same covariance matrix $\sum$. $R = \sum_{i=1}^{n} X_i\,{}^tX_i$ will be the Wilks-Wishart matrix of the first n vectors. The Hotelling ratio will be $[\mathrm{tr}[\sum_{i=1}^{n} X_i\,{}^tX_i]^{-1}X\,{}^tX]^{1/2}$ ($(n\geq k)$ implies the existence of R^{-1}).

If U is a symmetrical matrix, the terms outside the principal diagonal being quoted $\frac{1}{2}u_{ij} = \frac{1}{2}u_{ji}$ the characteristic function of $R + X\,{}^tX$ and $\mathrm{tr}\,R^{-1}X\,{}^tX$ will be:

$$\phi(U,v) = E[\exp i\ \mathrm{tr}[U(R + X\,{}^tX) + vR^{-1}X\,{}^tX]$$

$$= \frac{1}{\left((2\pi)^k \det \Sigma\right)^{\frac{n+1}{2}}} \int_{\mathbb{R}^{nk}} \exp -\frac{1}{2}\mathrm{tr}\left\{[\textstyle\sum^{-1}-2iU] \sum_{i=1}^{n} X_i\,{}^tX_i\right\} d(X_i)$$

ISBN: 0-12-166650-6

$$\times \int_{\mathbb{R}^k} \exp - \frac{1}{2} \operatorname{tr}\left\{\Sigma^{-1} - 2iU - 2iv[\sum_{i=1}^{n} X_i\ {}^tX_i]^{-1}\right\} X\ {}^tX\ d(X)$$

$$= \frac{1}{(2\pi)^{nk/2}(\det \Sigma)^{\frac{n+1}{2}}} \int_{\mathbb{R}^{nk}} \exp - \frac{1}{2} \operatorname{tr}[\Sigma^{-1} - 2iU] \sum_{i=1}^{n} X_i\ {}^tX_i$$

$$\times [\det\left\{\Sigma^{-1} - 2iU - 2iv[\sum_{i=1}^{n} X_i {}^tX_i]^{-1}\right\}]^{-1/2}\ d(X_i).$$

Let us put $[\Sigma^{-1} - 2iU]^{1/2}\ X_i = Y_i$. So

$$\operatorname{tr}[\Sigma^{-1} - 2iU] \sum_{i=1}^{n} X_i {}^tX_i = \operatorname{tr} \sum_{i=1}^{n} Y_i {}^tY_i$$

$$\frac{D(X_i)}{D(Y_i)} = \det[\Sigma^{-1} - 2iU]^{-1/2}.$$

$$\det \left\{ \Sigma^{-1} - 2iU - 2iv[\sum_{i=1}^{n} X_i {}^tX_i]^{-1}\right\}$$

$$= \det[\Sigma^{-1} - 2iU]\det[I - 2iv[\sum_{i=1}^{n} Y_i {}^tY_i]^{-1}].$$

And then,

$$\phi(U,v) = \frac{1}{(2\pi)^{\frac{nk}{2}}} \frac{1}{(\det \Sigma)^{\frac{n+1}{2}}} \frac{1}{\det[\Sigma^{-1} - 2iU]^{\frac{n+1}{2}}}$$

$$\times \int_{\mathbb{R}^{nk}} \frac{\exp - \frac{1}{2} \operatorname{tr} \sum_{i=1}^{n} Y_i {}^tY_i}{\det[1 - 2iv[\sum_{i=1}^{n} Y_i {}^tY_i]^{-1}]^{1/2}}\ dY_i\ .$$

1. We have $\phi(U,v) = \phi(0,v)\phi(U,0)$ and $R + X\,{}^tX$ and $\operatorname{tr} R^{-1}X\,{}^tX$ are independent; $\phi(U,0) = \det[1 - 2i\,\Sigma U]^{-(n+1)/2}$, $R + X\,{}^tX$ is a Wilks-Wishart matrix on $n + 1$ vectors which is obvious.

2. The characteristic function of $\operatorname{tr} R^{-1}X\,{}^tX$ is

$$\frac{1}{(2\pi)^{\frac{nk}{2}}} \int_{\mathbb{R}^{nk}} \frac{\exp -\frac{1}{2} \operatorname{tr} \sum_{i=1}^{n} Y_i\,{}^tY_i}{\det[1 - 2iv[\sum_{i=1}^{n} Y_i\,{}^tY_i]^{-1}]^{1/2}}\, d(Y_i) = \phi(0,v).$$

To have $P[(\operatorname{tr} R^{-1}X\,{}^tX)^{1/2} < x] = P[\operatorname{tr} R^{-1}X\,{}^tX < x^2]$ we must take the Fourier transform

$$\lim_{T=\infty} \frac{1}{2\pi} \int_{-T}^{+T} [\frac{1 - e^{-ix^2v}}{iv}] \frac{1}{(2\pi)^{\frac{nk}{2}}} \int_{\mathbb{R}^{nk}} \frac{\exp -\frac{1}{2} \operatorname{tr} \sum_{i=1}^{n} Y_i\,{}^tY_i}{(\det[1-2iv[\sum_{i=1}^{n} Y_i\,{}^tY_i]^{-1})^{1/2}}$$

$$d(Y_i)]dv.$$

Using Fubini's theorem, we have this integral equal to

$$\frac{1}{(2\pi)^{\frac{nk}{2}}} \int_{\mathbb{R}^{nk}} \exp -\frac{1}{2} \operatorname{tr} \sum_{i=1}^{n} Y_i\,{}^tY_i d(Y_i) \lim_{T=\infty} \frac{1}{2\pi}$$

$$\int_{-T}^{+T} \frac{1}{\det[1-2iv[\sum_{i=1}^{n} Y_i\,{}^tY_i]^{-1}]^{1/2}} \frac{1-e^{-ix^2v}}{iv}\, dv$$

$\lim_{T=\infty} \frac{1}{2\pi} \int_{-T}^{+T} \frac{1}{\det[1-2iv[\sum_{i=1}^{n} Y_i\,{}^tY_i]^{-1}]^{1/2}} \frac{1-e^{-ix^2v}}{iv}\, dv$ is the probability that the trace of the matrix $\Xi\,{}^t\Xi$ is less than x^2, Ξ being a k-dimensional normal vector, with zero mean and $[\sum_{i=1}^{n} Y_i\,{}^tY_i]^{-1}$ covariance matrix. This probability will be

$$\frac{1}{(2\pi)^{k/2}} \int_{\mathrm{tr}\Xi\,{}^t\Xi<x^2} [\det \sum_{i=1}^{n} Y_i\,{}^tY_i]^{1/2} \exp - \frac{1}{2}\,\mathrm{tr}[\sum_{i=1}^{n} Y_i\,{}^tY_i]\Xi\,{}^t\Xi d(\Xi).$$

And so,

$$P[(\mathrm{tr}\ R^{-1}X\ {}^tX)^{1/2} < x] = \frac{1}{(2\pi)^{\frac{(n+1)k}{2}}} \int_{\mathrm{tr}\Xi\,{}^t\Xi<x^2} d(\Xi) \int_{\mathbb{R}^{nk}} [\det \sum_{i=1}^{n} Y_i\,{}^tY_i]^{1/2}$$

$$\times \exp - \frac{1}{2}\,\mathrm{tr}[1 + \Xi\,{}^t\Xi] \sum_{i=1}^{n} Y_i\,{}^tY_i d(Y_i).$$

Putting $[1 + \Xi\,{}^t\Xi]^{1/2}\ Y_i = Z_i$, the second integral becomes

$$\det[1 + \Xi\,{}^t\Xi]^{-(n+1)/2} \int_{\mathbb{R}^{nk}} [\det \sum_{i=1}^{n} Z_i\,{}^tZ_i]^{1/2} \exp - \frac{1}{2}\,\mathrm{tr} \sum_{i=1}^{n} Z_i\,{}^tZ_i\ dZ_i$$

and

$$P[[\mathrm{tr}\ R^{-1}X\ {}^tX]^{1/2} < x] = C \int_{\mathrm{tr}\Xi\,{}^t\Xi<x^2} \frac{d\Xi}{\det[1+\Xi\,{}^t\Xi]^{(n+1)/2}}\ .$$

Let us take ${}^t\Xi = (\Xi_1,\ldots,\Xi_k)$ and put $\mathrm{tr}\Xi\ \Xi^t = r^2$. It is easy to see that $\det[1 + \Xi\,{}^t\Xi] = 1 + \Xi_1^2+\ldots+\Xi_k^2 = 1 + r^2$. And in the k-dimensional space the differential element will be

$$r^{k-1}f(\theta_1,\ldots,\theta_{k-1})d\theta_1\ldots d\theta_{k-1}dr,$$

$\theta_1,\ldots,\theta_{k-1}$ being "angular" variables. Thus

$$P[[\mathrm{tr}\ R^{-1}X\ {}^tX]^{1/2} < x] = C' \int_0^x \frac{r^{k-1}dr}{(1+r^2)^{\frac{n+1}{2}}}$$

which is the well-known Hotelling's T distribution. In the case $k = 1$, we have Student's t-distribution.

II

All that follows is a mixture of the work of Paul Levy, Kolmogoroff and Hoeffding, three great mathematicians, and a little bit of personal work.

Calculation of eigenvalues of special matrices.

a) Let us call A_n a triangular $n \times n$ matrix all of whose terms a_{ij} are equal to 1 if $j \le i$ and 0 otherwise.

Let us calculate

$$D_n(z) = \det[I_n - z({}^tA_n - A_n)].$$

$$I_n - z({}^tA_n - A_n) = \begin{pmatrix} 1 & -z & \cdots & & -z \\ z & 1 & & & \\ & & 1 & & \\ \vdots & & & \ddots & \\ z & & & & 1 \end{pmatrix}.$$

After the two operations:

first subtracting all the elements of the second line from the first one and then subtracting all the elements of the second column from the first one, we have the following matrix whose determinant is equal to $D_n(z)$.

$$\begin{pmatrix} 2 & -(1+z) & 0 & \cdots & 0 \\ -(1-z) & 1 & -z & & -z \\ 0 & z & & & \\ \vdots & & & & \\ 0 & z & & & 1 \end{pmatrix}.$$

Expanding through the elements of the first line we get

$$D_n(z) = 2D_{n-1}(z) - (1-z^2)D_{n-2}(z)$$

which is a Fibonacci recurrent equation. The general solution will be $Ar_1^n + Br_2^n$, r_1 and r_2 being the roots of $r^2 - 2r + (1-z^2)$. So $r_1 = 1-z$ and $r_2 = 1+z$. D_1 being equal to 1, D_2 to $1+z^2$, we have $D_n(z) = \frac{1}{2}[(1+z)^n + (1-z)^n]$. D_n has a degree equal to n if n is even and $n-1$ if n is odd. The roots of $D_n(z)$ are $itg(k + \frac{1}{2})\frac{\pi}{n}$ (if n is equal to $2p$ or $2p+1$, k runs from $-p$ to $p-1$). And so

$$D_n(z) = \prod_{k=-p}^{p-1} \left[1 - \frac{z}{itg(k+ \frac{1}{2})\frac{\pi}{n}} \right] ,$$

n being equal to $2p$ or $2p+1$.

b) Now let us calculate

$$E_n(z) = \det[I_n-z(I_n- \frac{1}{n} M_n)({}^tA_n-A_n)]=\det[I_n-z({}^tA_n-A_n)+ \frac{zM_n}{n}({}^tA_n-A_n)],$$

M_n being a matrix all of whose elements are 1. And so in the matrix $M_n({}^tA_n-A_n)$, which is obviously of rank 1, the elements of the j-th column are all equal, each of them being $-(n-2j+1)$. If we expand $\det[I_n-z({}^tA_n-A_n) + \frac{z}{n} M_n({}^tA_n-A_n)]$ through the method of polynomial columns, we must take just one column in $\frac{z}{n} M_n({}^tA_n-A_n)$ and the other $n-1$ in $1_n-z({}^tA_n-A_n)$, otherwise the determinant of the matrix will be zero.

We then have n determinants to add to $\det[I_n-z({}^tA_n-A_n)]$, each of them being

$$
-\frac{n-2j+1}{n} \cdot \begin{pmatrix} 1 & -z & z & \cdots & -z \\ z & 1 & & & \\ & & z & & \\ \cdot & & & & -z \\ \cdot & & & & \\ \cdot & & & & \\ z & & z & & 1 \end{pmatrix}
$$

j-th column

If we subtract the j-th column from the first $j-1$ and add it to the last $(n-j)$, it is easy to see that the determinant of this matrix will be

$$
-\frac{n-2j+1}{n}(1-z)^{j-1}(1+z)^{n-j}\, z.
$$

And so we have by replacing $\det[I_n - z({}^tA_n - A_n)]$ by its value

$$
E_n(z) = \tfrac{1}{2}[(1+z)^n + (1-z)^n] - \frac{z}{n}\sum_{j=1}^{n}(n-2j+1)(1-z)^{j-1}(1+z)^{n-j}
$$

$$
E_n(z) = (1+z)^n\left[\frac{1}{2}\left[1 + \left(\frac{1-z}{1+z}\right)^n\right] - \frac{z}{n(1+z)}\sum_{j=1}^{n}(n-2j+1)\left(\frac{1-z}{1+z}\right)^{j-1}\right]
$$

We have: $\displaystyle\sum_{j=1}^{n}(n-2j+1)\left(\frac{1-z}{1+z}\right)^{j-1} = (n+1)\sum_{j=1}^{n}\left(\frac{1-z}{1+z}\right)^{j-1} - 2\sum_{j=1}^{n} j\left(\frac{1-z}{1+z}\right)^{j-1}$

and: $\displaystyle(n+1)\sum_{j=1}^{n}\left(\frac{1-z}{1+z}\right)^{j-1} = (n+1)\frac{1-\left(\frac{1-z}{1+z}\right)^n}{1-\left(\frac{1-z}{1+z}\right)}$

$$
,\quad \sum_{j=1}^{n} j\left(\frac{1-z}{1+z}\right)^{j-1} = \frac{d}{d\left(\frac{1-z}{1+z}\right)}\sum_{j=1}^{n}\left(\frac{1-z}{1+z}\right)^{j} = \frac{d}{d\left(\frac{1-z}{1+z}\right)}\left(\frac{1-z}{1+z}\right)\frac{1-\left(\frac{1-z}{1+z}\right)^n}{1-\left(\frac{1-z}{1+z}\right)}\ .
$$

All calculations done, we find $E_n(z) = \dfrac{(1+z)^n - (1-z)^n}{2nz}$.

The degree of $E_n(z)$ is $n-1$ if n is odd and $(n-2)$ if n is even. The roots of $E_n(z)$ are $i \operatorname{tg} k\frac{\pi}{n}$, $k \neq 0$ (if n is equal to $2p+1$ or $2p+2$ k runs from $-p$ to $+p$, 0 being excepted). And

$$E_n(z) = \prod_{\substack{k=-p \\ k\neq 0}}^{p} \left[1 - \frac{z}{i \operatorname{tg} k\frac{\pi}{n}}\right], \quad n \text{ being equal to } 2p+1 \text{ or } 2p+2.$$

c) $i({}^tA_n - A_n)$ is an Hermitian matrix. So it can be diagonalized through a unitarian matrix. Let us call D_n its diagonal form ${}^tU_n i({}^tA_n - A_n)U_n = D_n$ and let us consider the matrix ${}^tU_n M_n U_n = N_n$ and call n_k the diagonal elements of N_n. We have

$$D_n(z) = \det[I_n - z({}^tA_n - A_n)] = \det[I_n - \frac{z}{i} D_n]$$

and

$$E_n(z) = \det[I_n - z(I_n - \frac{1}{n} M_n)({}^tA_n - A_n)] = \det[I_n - \frac{z}{i} D_n + \frac{z}{ni} N_n D_n].$$

If we expand $\det[I_n - \frac{z}{i} D_n + \frac{z}{ni} N_n D_n]$ by the method of polynomial columns, as $N_n D_n$ is of rank 1 we have:

$$E_n(z) = D_n(z) + \sum_{j=1}^{n} \Delta_{n,j}(z),$$

$\Delta_{n,j}(z)$ being the determinant of a matrix $I - \frac{z}{i} D_n$ in which the j-th column of $D_n(z)$ is replaced by the j-th column of $\frac{z}{ni} N_n D_n$. And so if $n = 2p$

$$\prod_{\substack{k=-p+1\\k\neq 0}}^{p-1}\left(1-\frac{z}{i\ \mathrm{tg}\ k\frac{\pi}{2p}}\right)$$

$$=\prod_{k=-p}^{p-1}\left(1-\frac{z}{i\ \mathrm{tg}(k+\frac{1}{2})\frac{\pi}{2p}}\right)\left[1+\sum_{k=-p}^{p-1}\frac{\frac{z}{ni}\ \frac{n_k}{\mathrm{tg}(k+\frac{1}{2})\frac{\pi}{2p}}}{1-\frac{z}{i\ \mathrm{tg}(k+\frac{1}{2})\frac{\pi}{2p}}}\right]$$

Now

$$E_{2p}[i\ \mathrm{tg}(r+\tfrac{1}{2})\tfrac{\pi}{2p}]=\frac{n_r}{2p}\prod_{\substack{k=-p\\k\neq r}}^{p-1}\left(1-\frac{i\ \mathrm{tg}(r+\frac{1}{2})\frac{\pi}{2p}}{i\ \mathrm{tg}(k+\frac{1}{2})\frac{\pi}{2p}}\right)$$

$$\frac{d}{dz}D_{2p}\left(i\ \mathrm{tg}(r+\tfrac{1}{2})\tfrac{\pi}{2p}\right)=-\frac{1}{i\ \mathrm{tg}(r+\frac{1}{2})\frac{\pi}{2p}}\prod_{\substack{k=-p\\k\neq r}}^{p-1}\left(1-\frac{i\ \mathrm{tg}(r+\frac{1}{2})\frac{\pi}{2p}}{i\ \mathrm{tg}(k+\frac{1}{2})\frac{\pi}{2p}}\right)$$

$$\frac{d}{dz}D_{2p}\left(i\ \mathrm{tg}(r+\tfrac{1}{2})\tfrac{\pi}{2p}\right)=\frac{2p}{2}\ \frac{e^{i(r+\frac{1}{2})\frac{\pi}{2p}(2p-1)}-e^{-i(r+\frac{1}{2})\frac{\pi}{2p}(2p-1)}}{\left(\cos(r+\frac{1}{2})\frac{\pi}{2p}\right)^{2p-1}}\ .$$

$$\frac{d}{dz}D_{2p}\left(i\ \mathrm{tg}(r+\tfrac{1}{2})\tfrac{\pi}{2p}\right)=\frac{2p}{2}\ \frac{(-1)^r 2i\ \cos(r+\frac{1}{2})\frac{\pi}{2p}}{[\cos(r+\frac{1}{2})\frac{\pi}{2p}]^{2p-1}}$$

$$=(-1)^r\ 2pi\ \frac{1}{[\cos(r+\frac{1}{2})\frac{\pi}{2p}]^{2p-2}}$$

and

$$\frac{n_r}{2p}=-\frac{E_{2p}\left(i\ \mathrm{tg}(r+\frac{1}{2})\frac{\pi}{2p}\right)}{i[\mathrm{tg}(r+\frac{1}{2})\frac{\pi}{2p}]\ \frac{(-1)^r 2pi}{\left(\cos(r+\frac{1}{2})\frac{\pi}{2p}\right)^{2p-2}}}$$

$$E_{2p}\left(i\ \mathrm{tg}(r+\tfrac{1}{2})\tfrac{\pi}{2p}\right) = \frac{e^{i(r+\frac{1}{2})\pi} - e^{-i(r+\frac{1}{2})\pi}}{2p[\cos(r+\frac{1}{2})\frac{\pi}{2p}]^{2p}\ 2i\ \mathrm{tg}(r+\frac{1}{2})\frac{\pi}{2p}}$$

$$= \frac{(-1)^r}{2p[\cos(r+\frac{1}{2})\frac{\pi}{2p}]^{2p}\ \mathrm{tg}(r+\frac{1}{2})\frac{\pi}{2p}}$$

which gives

$$\frac{n_r}{2p} = \frac{1}{(2p)^2[\sin(r+\frac{1}{2})\frac{\pi}{2p}]^2} \quad . \tag{A}$$

We have

$$\sum_{r=-p}^{p-1} \frac{n_r}{2p} = \frac{1}{2p}\,\mathrm{tr}\ N_{2p} = \frac{1}{2p}\,\mathrm{tr}\ M_{2p} = 1. \tag{B}$$

So

$$\sum_{r=-p}^{p-1} \frac{1}{(2p)^2[\sin(r+\frac{1}{2})\frac{\pi}{2p}]^2} = 1$$

with analogous formulas if n is odd.

It is easy to see that:

$$\lim D_n(\tfrac{z}{n}) = \mathrm{Ch}\ z \quad \text{and} \quad \lim E_n(\tfrac{z}{n}) = \frac{\mathrm{Sh}\ z}{z} \quad .$$

Calling $A^{\otimes s}$ the s-th tensorial power of the matrix A we have:

$$D_n^{(s)}(z) = \det[I_n^{\otimes s} - z({}^tA_n - A_n)^{\otimes s}]$$

$$= \prod_{k_1=-p}^{p-1} \cdots \prod_{k_s=-p}^{p-1} \left[1 - \frac{z}{i\ \mathrm{tg}(k_1+\frac{1}{2})\frac{\pi}{n} \cdots i\ \mathrm{tg}(k_s+\frac{1}{2})\frac{\pi}{n}}\right]$$

n being equal to $2p$ or $2p+1$. And it is easy to see that

$$\lim_{n\to\infty} D_n^{(s)}\left(\frac{z}{n^s}\right) = \lim_{N\to\infty} \prod_{k_1=-N}^{+N} \cdots \prod_{k_s=-N}^{+N} \left(1 - \frac{z}{i(k_1+\frac{1}{2})\pi \ldots i(k_s+\frac{1}{2})\pi}\right)$$

$$= \left[\prod_{k_1=0}^{+\infty} \cdots \prod_{k_s=0}^{+\infty} \left[1+(-1)^{s+1}\frac{z^2}{(k_1+\frac{1}{2})^2\pi^2 \ldots (k_s+\frac{1}{2})^2\pi^2}\right]\right]^{2^{s-1}} .$$

We shall call it $D^{(s)}(z)$. Of course, $D^{(1)}(z) = \text{Ch}\, z$.

Now put

$$E_n^{(s)}(z) = \det[I_n^{\otimes s} - z\left(I_n^{\otimes s} - (\tfrac{1}{n} M_n)^{\otimes s}\right)({}^tA_n - A_n)^{\otimes s}].$$

If $n = 2p$ (and a similar formula for n odd) we have

$$E_{2p}^{(s)}(z) = \prod_{k_1=-p}^{p-1} \cdots \prod_{k_s=-p}^{p-1} \left(1 - \frac{z}{i\ \text{tg}(k_1+\frac{1}{2})\frac{\pi}{2p} \ldots i\ \text{tg}(k_s+\frac{1}{2})\frac{\pi}{2p}}\right)$$

$$\cdot \left[1 + \sum_{k_1=-p}^{p-1} \cdots \sum_{k_s=-p}^{p-1} \frac{\dfrac{z}{(2p)^s}\ \dfrac{n_{k_1} \ldots n_{k_s}}{i\ \text{tg}(k_1+\frac{1}{2})\frac{\pi}{2p} \ldots i\ \text{tg}(k_s+\frac{1}{2})\frac{\pi}{2p}}}{1 - \dfrac{z}{i\ \text{tg}(k_1+\frac{1}{2})\frac{\pi}{2p} \ldots i\ \text{tg}(k_s+\frac{1}{2})\frac{\pi}{2p}}}\right] .$$

After (B) we have

$$1 = \sum_{r=-p}^{p-1} \frac{n_r}{2p} = \left(\sum_{r=-p}^{p-1} \frac{n_r}{2p}\right)^s$$

and so the bracket can be written

$$\sum_{k_1=-p}^{p-1} \cdots \sum_{k_s=-p}^{p-1} \frac{\frac{n_{k_1}}{2p} \cdots \frac{n_{k_s}}{2p}}{1 - \frac{z}{i\ \mathrm{tg}(k_1 + \frac{1}{2})\frac{\pi}{2p} \cdots i\ \mathrm{tg}(k_s + \frac{1}{2})\frac{\pi}{2p}}}$$

and after (A):

$$\sum_{k_1=-p}^{p-1} \cdots \sum_{k_s=-p}^{p-1} \frac{\frac{1}{\left(2p \sin(k_1 + \frac{1}{2})\frac{\pi}{2p}\right)^2 \cdots \left(2p \sin(k_s + \frac{1}{2})\frac{\pi}{2p}\right)^2}}{1 - \frac{z}{i\ \mathrm{tg}(k_1 + \frac{1}{2})\frac{\pi}{2p} \cdots i\ \mathrm{tg}(k_s + \frac{1}{2})\frac{\pi}{2p}}} .$$

If we put $\frac{z}{(2p)^s}$ instead of z in this expression we have as a limit, when $n = 2p$ increases indefinitely

$$\lim_{N\to\infty} \sum_{k_1=-N}^{N} \cdots \sum_{k_s=-N}^{N} \frac{\frac{1}{(k_1 + \frac{1}{2})^2\pi^2 \cdots (k_s + \frac{1}{2})^2\pi^2}}{1 - \frac{z}{i(k_1 + \frac{1}{2})\pi \cdots i(k_s + \frac{1}{2})\pi}}$$

which is equal to

$$2^{s-1} \sum_{k_1=0}^{+\infty} \cdots \sum_{k_s=0}^{+\infty} \frac{\frac{2}{(k_1 + \frac{1}{2})^2\pi^2 \cdots (k_s + \frac{1}{2})^2\pi^2}}{1 + (-1)^{s+1} \frac{z^2}{(k_1 + \frac{1}{2})^2\pi^2 \cdots (k_s + \frac{1}{2})^2\pi^2}}$$

$$= 2^{s-1} \frac{(-1)^{s+1}}{z} \sum_{k_1=0}^{\infty} \cdots \sum_{k_s=0}^{\infty} \frac{d}{dz}\left[\log 1+(-1)^{s+1} \frac{z^2}{(k_1+\frac{1}{2})^2\pi^2 \cdots (k_s+\frac{1}{2})^2\pi^2}\right]$$

$$= 2^{s-1} \frac{(-1)^{s+1}}{z} \frac{d}{dz} \log \sum_{k_1=0}^{\infty} \cdots \sum_{k_s=0}^{\infty} \left(1+(-1)^{s+1} \frac{z^2}{(k_1+\frac{1}{2})^2\pi^2 \cdots (k_2+\frac{1}{2})^2\pi^2}\right)$$

$$= \frac{(-1)^{s+1}}{z} \frac{d}{dz} \log D^{(s)}(z) = \frac{(-1)^{s+1}}{z} \frac{\frac{d}{dz} D^{(s)}(z)}{D^{(s)}(z)}$$

and so

$$\lim E_{2p}^{(s)}\left(\frac{z}{(2p)^s}\right) = E^{(s)}(z) = \frac{(-1)^{s+1}}{z} \frac{d}{dz} D^{(s)}(z) .$$

Obviously,

$$E^{(1)}(z) = \frac{1}{z}\frac{d}{dz} \text{Ch } z = \frac{\text{Sh } z}{z} .$$

III

Extension of the Preceeding Results

It is easy to see that all the preceeding results can be extended in the following way:

(a) $\lim_{n\to\infty} \det[I_n - zC_n({}^tA_n - A_n)] = D^{(1)}(z).$

C_n being a diagonal matrix whose all elements are positive and trace equal to 1. If $\Delta F(\frac{i}{n})$ is the ith element on the diagonal with $F(1) - F(0) = 1$ the matrix $C_n({}^tA_n - A_n)$ tends towards the kernel

$N(x,y) = dF(x)$ for $x < y$

$- dF(x)$ for $x > y$.

And the eigenvalues of $C_n({}^tA_n - A_n)$ tends towards the eigenvalues of the integral equation:

$$\phi(y) = \lambda \left[\int_0^y \phi(x)dF(x) - \int_y^1 \phi(x)dF(x) \right] . \quad (C)$$

If F is continuous $\phi(y)$ is continuous and derivable with respect to $F(y)$. So we have $\frac{d\phi(y)}{dF(y)} = 2\lambda\phi$ whose solution is $K \exp 2\lambda F$. Putting in (C) we get $\lambda = i(k + \frac{1}{2})\pi$, k running from $-\infty$ to $+\infty$. Which completes the proof that:

$$\lim \det[I_n - zC_n({}^tA_n - A_n)] = D^{(1)}(z) = \operatorname{Ch} z.$$

(b) Now we have:

$$\lim_{n\to\infty} \det[I_n - z(C_n - C_nM_nC_n)({}^tA_n - A_n)] = E^{(1)}(z).$$

If we call Γ_n a vector whose n coordinates are $\Delta F(\frac{i}{n})$ the matrix $C_nM_nC_n$ is $\Gamma_n \cdot {}^t\Gamma_n$ and so when n increases indefinitely the eigenvalues of $(C_n - C_nM_nC_n)({}^tA_n - A_n) = (C_n - \Gamma_n{}^t\Gamma_n)({}^tA_n - A_n)$ will tend towards that of the integral equation:

$$\phi(y) = \lambda\left[\int_0^y \phi(x)dF(x) - \int_y^1 \phi(x)dF(x) - \left(\int_0^1 \phi(x)dF(x)(2F(y)-1\right)\right] \quad \text{(D)}$$

If <u>F is continuous</u> we can differentiate with respect to $F(y)$; ϕ must be a solution of $\frac{d\phi}{dF(y)} = 2\lambda[\phi - \int_0^1 \phi dF]$ and so $\frac{d^2\phi}{dF^2} = 2\lambda \frac{d\phi}{dF} \cdot \phi$ must be equal to $\frac{C}{2\lambda} \exp 2\lambda F + D$ and putting in (D) we get: $\lambda = k\pi i$, k running from $-\infty$ to $+\infty$ and being different from 0. So:

$$\lim_{n\to\infty} \det[I_n - z(C_n - C_nM_nC_n)({}^tA_n - A_n)] = E^{(1)}(z) = \frac{\operatorname{Sh} z}{z}.$$

(c) If $C_n({}^tA_n - A)$ is diagonalized by the means of a matrix T_n (which is no longer unitary) $T_n^{-1}C_n({}^tA_n - A_n)T_n = B_n$ we shall put: $T_n^{-1}C_nM_nT_n = P_n$ which is of course of rank 1 as M_n.

We have:

$$\det[I_n - z(C_n - C_n M_n C_n)({}^tA_n - A_n)] = \det[I_n - zB_n + zP_n B_n].$$

Exactly as in II (c), calling $p_r^{(n)}$ the rth term on the diagonal of P_n, we have $\lim\limits_{n\to\infty} p_r^{(n)} = \dfrac{1}{(r+\frac{1}{2})^2\pi^2}$. We have to calculate

$\dfrac{\mathrm{Sh}\left(i(r+\frac{1}{2})\pi\right)}{i(r+\frac{1}{2})\pi}$ and $\dfrac{d}{dz}\,\mathrm{Ch}\left(i(r+\frac{1}{2})\pi\right)$ and we use the fact that if on a close curve an analytic function tends towards an analytic function inside the curve, the derivative of the first one tends towards the derivative of the second. And as the trace of P_n is equal to the trace of $C_n M_n$ that is:

$$\Delta F(\tfrac{1}{n}) + \Delta F(\tfrac{2}{n}) + \ldots + \Delta F(\tfrac{n}{n}) = 1; \quad \sum_{r=-\infty}^{+\infty} \frac{1}{(r+\frac{1}{2})^2\,\pi^2} = 1$$

(d) The extension of I (d) follows immediately and is based on the preceding equality; we have:

$$\lim_{n\to\infty} \det[I_n^{\otimes s} - zC_n^{\otimes s}({}^tA_n - A_n)^{\otimes s}] = D^{(s)}(z) \quad \text{exactly as}$$

$\lim\limits_{n\to\infty} \det[I_n - zC_n({}^tA_n - A_n)] = D^{(1)}(z)$. The proof is obvious as the eigenvalues of $C_n^{\otimes s}({}^tA_n - A_n)^{\otimes s} = [C_n({}^tA_n - A_n)]^{\otimes s}$ are every product s to s of the eigenvalues of $C_n({}^tA_n - A_n)$

$$\lim_{n\to\infty} [I_n^{\otimes s} - z(C_n^{\otimes s} - (C_n M_n C_n)^{\otimes s})({}^tA_n - A_n)^{\otimes s}] =$$

$$D^{(s)}(z)\, 2^{s-1} \sum_{k_1=0}^{+\infty} \cdots \sum_{k_s=0}^{+\infty} \frac{\dfrac{2}{(k_1+\frac{1}{2})^2\pi^2 \cdots (k_s+\frac{1}{2})\pi^2}}{1+(-1)^{s+1} \dfrac{z^2}{(k_1+\frac{1}{2})^2\pi^2 \cdots (k_s+\frac{1}{2})^2\pi^2}}$$

which is equal, as it was shown, to:

$$\frac{(-1)^{s+1}}{z} \frac{d}{dz} D^{(s)}(z).$$

The calculations are exactly the same as in I (d).

Remark. The results of (d) would be exactly the same if instead of $C_n^{\otimes s}$ and $(C_n M_n C_n)^{\otimes s}$ we had $\bigotimes_{i=1}^{s} C_n^{(i)}$ and $\bigotimes_{i=1}^{s} C_n^{(i)} M_n C_n^{(i)}$.

IV.

Application to a problem of Von Mises Smirnoff

(a) Let us consider a cumulated histogram of frequencies on a s dimensional variable $H_p(x_1,\ldots,x_s)$ whose probability law is $F_1(x_1)\ldots F_s(x_s)$ (independence of the coordinates, the F_i being continuous). Let us consider, p being the sample size the n^s-dimensional variables which are the increments of $\sqrt{p}[H_p - F_1 \ldots F_s]$ each coordinate being divided in n segments

$$\sqrt{p}[\Delta_{x_1\ldots x_s} H_p(x_1,\ldots,x_s) - \Delta F_1(x_1)\ldots\Delta F_s(x_s)] = \sqrt{p}\ \Delta\ J_p(x_1,\ldots,x_s)$$

When p increases indefinitely this n^s dimensional variable has a law which tends towards a normal law with mean zero and covariance matrix

$$\bigotimes_{i=1}^{s} C_n^{(i)} - \bigotimes_{i=1}^{s} C_n^{(i)} M_n C_n^{(i)}$$

$C_n^{(i)}$ being a diagonal matrix whose all terms are $\Delta F_i(x_i^{(1)}),\ldots,$ $\Delta F_i(x_i^{(n)})$. Calling $J_p(x_1 \ldots x_s)$ the difference $H_p(x_1 \ldots x_s) -$ $F_1(x_1)\ldots F_s(x_s)$ and considering two lots of p results we have $J_p^{(1)}(x_1^{(1)} \ldots x_s^{(1)})$ and $J_p^{(2)}(x_1^{(2)} \ldots x_s^{(2)})$. Now we can write the integral

$$L = p\int_{\mathbb{R}^s \mathbb{R}^s} \phi(x_1^{(1)},\ldots x_s^{(1)}, x_1^{(2)},\ldots, x_s^{(2)})\, dJ_p^{(1)}(x_1^{(1)} \ldots x_s^{(1)})\, dJ_p^{(2)}(x_1^{(2)} \ldots x_s^{(2)})$$

$\phi(x_1^{(1)},\ldots,x_s^{(1)},\ldots,x_s^{(1)},\ x_1^{(2)},\ldots,x_s^{(2)})$ being equal to the product:

$$d(x_i^{(1)}, x_1^{(2)}) \ \ldots\ d(x_s^{(1)}, x_s^{(2)})$$

$d(x_i^{(1)}, x_i^{(2)})$ being equal to $+1$ if $x_i^{(2)} > x_i^{(1)}$ and -1 if $x_i^{(2)} <$ $x_i^{(1)}$. If $s = 1$ it is easy to see that the integral L is equal to the twice the area swept by the vector $(\sqrt{p}\, J_p^{(1)}(x_1^{(1)}),$ $\sqrt{p}\, J_p^{(2)}(x_1^{(2)}))$.

The Riemanian approximation of L through a quadratic and symmetric form will be obtained by the saturation of the matrix:

$$\frac{1}{2}\begin{pmatrix}0 & 1\\ 1 & 0\end{pmatrix} \otimes ({}^tA_n - A_n)^{\otimes s} \quad \text{if } s \text{ is even}$$

and

$$\frac{1}{2}\begin{pmatrix}0 & 1\\ -1 & 0\end{pmatrix} \otimes ({}^tA_n - A_n)^{\otimes s} \quad \text{if } s \text{ is odd.}$$

And so the characteristic function of $L = E(\exp iu\, L)$ will tend when p increases indefinitely towards the limit of

$$E\left[\exp iu\ {}^tX\ \frac{1}{2}\begin{pmatrix}0 & 1\\ 1 & 0\end{pmatrix} \otimes ({}^tA_n - A_n)^{\otimes s}\ X\right] \quad \text{or}$$

$$E\left[\exp iu\ {}^tX\ \frac{1}{2}\begin{pmatrix}0 & 1\\ -1 & 0\end{pmatrix} \otimes ({}^tA_n - A_n)^{\otimes s} X\right]$$

according to the parity of s when n increases indefinitely, X being the increments of $J^{(1)}(x_1^{(1)} \ldots x_s^{(1)})$ and $J^{(2)}(x_1^{(2)} \ldots x_s^{(2)})$ in lexicographic order (of course $2n^s$ components).

These increments will have, as seen in the beginning of the paragraph, a covariance matrix equal to:

$$\begin{pmatrix}1 & 0\\ 0 & 1\end{pmatrix} \otimes \left[\bigotimes_{i=1}^{s} C_n^{(i)} - \bigotimes_{i=1}^{s} C_n^{(i)} M_n C_n^{(i)}\right] .$$

And so $\lim_{p\to\infty} E(\exp iu\, L)$ is equal to the limit of

$$\det\left\{I_2 \otimes I_n - 2iu\,\frac{1}{2}\begin{bmatrix} 0 & 1 \\ (-1)^s & 0 \end{bmatrix} \otimes ({}^tA_n - A_n)^{\otimes s} \cdot \begin{bmatrix} 1 & 0 \\ 0 & 1 \end{bmatrix} \otimes \left[\bigotimes_{i=1}^{s} C_n^{(i)} - \bigotimes_{i=1}^{s} C_n^{(i)} M_n C_n^{(i)}\right]\right\}^{-\frac{1}{2}}$$

As we saw:

$$\det\left[I_n^{\otimes s} - iu\left({}^tA_n - A_n\right)^{\otimes s} \cdot \left(\bigotimes_{i=1}^{s} C_n^{(i)} - \bigotimes_{i=1}^{s} C_n^{(i)} M_n C_n^{(i)}\right)\right]$$

tends when n increases indefinitely towards $(-1)^s \frac{1}{u} \frac{dD^{(s)}(iu)}{du}$.

If s is even it is easy to see that, because $\begin{pmatrix} 0 & 1 \\ 1 & 0 \end{pmatrix}$ can be diagonalized and has a diagonal form equal to $\begin{pmatrix} 1 & 0 \\ 0 & -1 \end{pmatrix}$ that

$$\lim_{p\to\infty} E[\exp iu\,L) = \left[\frac{1}{u} \frac{dD^{(s)}(iu)}{du}\right]^{-1},$$ $D^{(s)}(z)$ being an even function. If s is odd, as the diagonal form of $\begin{pmatrix} 0 & 1 \\ -1 & 0 \end{pmatrix}$ is $\begin{pmatrix} i & 0 \\ 0 & -i \end{pmatrix}$ we have:

$$\lim_{p\to\infty} E[\exp iu\,L] = \left[\frac{1}{u} \frac{dD^{(s)}(u)}{du}\right]^{-1}.$$

If s is even:

$$D^{(s)}(iu) = \left[\prod_{k_1=0}^{\infty} \cdots \prod_{k_n=0}^{\infty} \left[1 + \frac{u^2}{(k_1+\frac{1}{2})^2\pi^2 \cdots (k_s+\frac{1}{2})^2\pi^2}\right]\right]^{2^{s-1}}$$

if s is odd:

$$D^{(s)}(u) = \left[\prod_{k_1=0}^{\infty} \cdots \prod_{k_s=0}^{\infty} \left[1 + \frac{u^2}{(k_1+\frac{1}{2})^2\pi^2 \cdots (k_s+\frac{1}{2})^2\pi^2} \right] \right]^{2^{s-1}} .$$

Let us call the second number of this two equalities $\Delta^{(s)}(u)$. So we have the general formula:

$$\lim_{p\to\infty} E(\exp iu\, L) = \left[\frac{1}{u} \frac{d}{du} \Delta^{(s)}(u) \right]^{-1} .$$

This result can be extended in the same manner as Hoeffding (1948).

BIBLIOGRAPHY

Blum, J.R., Kiefer, J. and Rosenblatt, M. Annals of Mathematical Statistics, 32, 1961, p. 485-497.

Dugué, D. Multivariate analysis. Dayton (Ohio) Colloquium 1968, p. 289-301.

Dugué, D. Comptes Rendus de l'Académie des Sciences, 281, 1975, p. 1103-1104.

Hoeffding, W. Annals of Mathematical Statistics, 19, 1948, p. 374-401.

Levy, Paul. Processus Stochastiques et Mouvement Brownien, 2-édition, 1965, p. 329-333.

A CHERNOFF-SAVAGE THEOREM FOR CORRELATION RANK STATISTICS WITH APPLICATIONS TO SEQUENTIAL TESTING

N. Bönner
U. Müller-Funk
H. Witting

Institut für Mathematische Stochastik
Universität Freiburg

The remainder term in the Chernoff-Savage representation of correlation rank statistics is shown to be of order $n^{-1/2-\gamma}$, $\gamma = \gamma(\kappa) > 0$, with probability $1 - n^{-\kappa}$ for all $\kappa > 0$ uniform over the class of all continuous bivariate distribution functions. At the crucial point of the proof a linear programming approach is employed. The theorem gives various weak convergence results and a law of the iterated logarithm. Moreover, a class of nonparametric SPRT-type tests for testing independence is studied. Asymptotic OC- and ASN-curves are derived. These tests are justified by asymptotic comparison with the corresponding SPRT.

I. INTRODUCTION AND SUMMARY

Let $(X_1,Y_1),(X_2,Y_2),\ldots$ be a sequence of independent identically distributed random variables (i.i.d. r.v.) the common distribution function (d.f.) H of which is supposed to belong to the class $\mathcal{H}$ of all continuous bivariate d.f. The marginals of H are designated as F and G. The empirical d.f. based on a sample of size n are denoted by H_n, F_n, and G_n, respectively. We shall also use the symbols $H_n^* = \frac{n}{n+1} H_n$, $F_n^* = \frac{n}{n+1} F_n$, and $G_n^* = \frac{n}{n+1} G_n$ for short.

For testing the hypothesis of independence the following sequential rank test was proposed by Sen and Ghosh (1974). Let R_{ni} (and S_{ni}) be the rank of X_i (and Y_i) among $X_1,\ldots,X_n$ (and $Y_1,\ldots,Y_n$) for i =

ISBN: 0-12-166650-6

$1,\ldots,n$ and let the linear correlation rank statistic be defined by

$$T_n := T_{n\,g_1 g_2} := \frac{1}{n} \sum_{i=1}^{n} t_1 \left(\frac{R_{ni}}{n+1}\right) g_2 \left(\frac{S_{ni}}{n+1}\right), \tag{1.1}$$

where g_1 and g_2 are some specified score functions. Suppose we want to test A_o: $\mu = 0$ (independence) against A_1: $\mu = \Delta > 0$ (positive dependence) with given error probabilities α and β in a one-dimensional subclass of continuous distributions which is reasonably parametrized. The sequential procedure of Sen and Ghosh (1974) works as follows: Start with an initial sample of size $n_o(\Delta)$ moderately large, and continue drawing observations one by one as long as

$$b < n\Delta\left(T_n - \frac{\Delta}{2}\right) < a \tag{1.2}$$

where $a := \log((1-\beta)/\alpha)$ and $b := \log(\beta/(1-\alpha))$. If $N(\Delta)$ is the first time at which (1.2) is violated, then accept A_o or A_1 according as $n\Delta(T_n - \frac{\Delta}{2})$ is $\leqq b$ or $\geqq a$. By virtue of a certain Wiener process approximation (as $\Delta \to 0$) Sen and Ghosh (1974) heuristically derived OC- and ASN-curves.

In this paper the sequential procedure just described is investigated in more detail and extended in various directions. As a first result we shall get in theorem 3.1 that the test based on (1.1) is unbiased against the subclass of positively regression dependent distributions, provided g_1 and g_2 are nondecreasing. A proper justification, however, can only be given by asymptotic considerations. To do so we shall start in theorem 3.2 with a suitable Wiener process approximation (as $\Delta \to 0$) for the rank statistic process under a one-parameter subclass $\{H_\Delta:\ 0 \leqq \Delta \leqq \Delta_o\} \subset \mathcal{H}$. The standardizing constants are to behave in a specific way for $\Delta \to 0$. This rank statistic process is built upon the whole sequence $(T_n)_{n \geqq 1}$ and not only upon the tail which makes it possible to do away with the initial sample needed by Sen and Ghosh (1974). Our invariance principle is a limit theorem on continuous convergence (with respect to families of d.f.). The need for uniform invariance principles has also been emphasized in a recent paper by Lai (1978). His results concern the two-sample case among others and differ from the type of functional limit theorem given here in formulation rather than in content; our formulation lays more emphasis on representing the asymptotic shift in order to derive explicit expressions for the asymptotic quantities we are looking for.

Analytically we shall obtain certain limiting expressions for the asymptotic OC- and ASN-functions. At this point it is not obvious that these purely probabilistic results apply to classes of distributions the elements of which are positively (or negatively) dependent. Here we adopt a class of nonparametric alternatives introduced in the non-sequential case by Behnen (1972). This author showed that for every pair (h_1,h_2) of sufficiently smooth and non-decreasing score functions there corresponds a subclass $\mathcal{H}_{h_1h_2}$ of alternatives which are positive quadrant dependent. It is a consequence of our corollary 3.4, that the sequential procedure for testing $\mu = 0$ against $\mu = \Delta$ in $\mathcal{H}_{h_1h_2}$ based on the rank statistics $T_{n\ h_1h_2}$ with stopping rule (1.2) is asymptotically optimal in the class if all sequential tests (not necessarily based on ranks), for which the kind of Wiener process approximation indicated above can be carried out. This optimality is in the sense that these tests have the same asymptotic OC- and ASN-function as the corresponding SPRT, whose optimality is well known.

In this context the concept of Pitman efficiency is easily extended to the sequential case. Note that a sequential version of the classical Pitman-Noether-theorem has recently been proved by Lai (1978). In (3.14) explicit expressions are given for the asymptotic relative efficiency of an arbitrary sequential rank test based on $T_{n\ g_1g_2}$ compared with the optimal one, when the underlying distribution H_Δ

belongs to $\mathcal{H}_{h_1 h_2}$ and $\Delta \to 0$. It turns out that for sequential tests the same expressions hold true as for the non-sequential ones. As a consequence, the bounds for the asymptotic relative efficiency obtained by Behnen (1972) are still valid in the present context.

Our theorems are also useful for the study of restricted sequential procedures, c.f. Sen (1978), or for the construction of sequential bounded length confidence intervals in the spirit of Chow and Robbins (1965) as it will be sketched at the end of the paper.

All these derivations are based on the Chernoff-Savage type theorem 2.4, which gives a $O(n^{-\frac{1}{2}-\eta})$ P_H-a.s.-statement uniformly in $H \in \mathcal{H}$ rather than the usual $o_p(n^{-1/2})$-statement. Moreover following Bhuchongkul's (1964) first Chernoff-Savage type theorem for the correlation case interest has been concentrated on relaxing the regularity conditions which are required for the score functions, cf. Ruymgaart, Shorack, van Zwet (1972) and Ruymgaart (1974). However it is the pair (g_1,g_2) of score functions that the statistician can choose freely whereas the underlying distribution evades his influence. Therefore, we are more interested in results not depending on the underlying distribution $H \in \mathcal{H}$ even at the cost of somewhat more stringent regularity conditions concerning the score functions. In theorem 2.4 we shall sharpen the Chernoff-Savage type statement but we shall keep Bhuchongkul's original assumptions.

A first theorem of this kind was proved by N. Bönner (1976) but his proof required the existence of a bounded density of H with respect to the product measure $F \otimes G$ of the marginal distributions in order to get non-stochastic upper bounds for integrals like $\iint f_1(F_n) f_2(G_n) dH_n$ or $\iint f_1(F) f_2(G) dH$ by factorization and by a "(discrete) change of variables". This assumption was motivated by the situation in the two-sample or in the one-sample symmetry case. In both cases the empirical d.f. in the integrand of the Chernoff-Savage representation dominates the empirical d.f. with respect to which the integration is carried out. The same strong stochastic ordering is also valid for the corresponding theoretical d.f. and hence the existence of a density poses no problem. But in our case continuous distributions may occur which do not possess a density with respect to the product of the marginals, for instance any continuous distribution which concentrates on the diagonal.

As the proof of theorem 2.4 shows the comparison of the underlying distribution H with the product measure $F\otimes G$ can be done without assuming bounded densities by comparing probabilities over certain rectangles.

Finally we should mention an unpublished paper by Tran (1978), in which a somewhat intermediate result is stated. But his proof does not overcome the difficulties discussed above.

II. A CHERNOFF-SAVAGE THEOREM

In addition to the terminology introduced at the beginning some further notation has to be adopted. If dependence on a certain d.f. H is to be stressed, we shall write $P_H(\cdot)$ etc. Notationally, a d.f. and the corresponding measure on the real line will be identified. In the ordinary way, we define the inverse of a d.f. as follows

$$F^{-1}(t): = \min\{x \in \mathbb{R}^1: \ F(x) \geq t\} \quad \text{for} \quad 0 < t < 1,$$

$$F^{-1}(0): = -\infty, \quad F^{-1}(1): = \infty.$$

$U_{n[1]} < \ldots < U_{n[n]}$ denotes an ordered random sample of size n which is drawn from the rectangular distribution. As usual, Φ stands for the d.f. of the standard normal law and " $\underset{\mathcal{D}}{\rightarrow}$" is to mean convergence in distribution. Frequent use will be made of a generic constant K which is allowed to vary from expression to expression. {x} denotes the smallest integer larger than or equal to some real number x ("waiter symbol"). $[x]$ is used in its familiar meaning. In the sequel, we shall introduce rectangles J_n, J_{nij} etc. understanding that all sides are parallel to the axes. Most score functions g_i to be considered here, $g_i:]0,1[\to \mathbb{R}^1$ $(i = 1,2)$, are assumed to be twice continuously differentiable and to satisfy the Chernoff-Savage type growth condition (C) due to Bhuchongkul (1964), i.e.

There exists some $K > 0$ so that for $i = 1,2$: (C)

$$|g_i''(t)| \leqq K(t(1-t))^{-2}, \quad \forall\, 0 < t < 1. \tag{2.1}$$

It is well-known, c.f. also Lai (1975a), that (C) implies

$$|g_i(t)| \leqq -K \log(t(1-t)), \tag{2.2}$$

$$|g_i(t)| \leqq K(t(1-t))^{-\eta}, \quad \forall\, \eta > 0, \tag{2.3}$$

$$|g_i'(t)| \leqq K(t(1-t))^{-1}, \tag{2.4}$$

where $i = 1,2$; $0 < t < 1$. We shall only make use of (2.3), but never of (2.2). To simplify the notation, we introduce the function $r(t) := t(1-t)$ for $0 < t < 1$. Of the examples satisfying condition (C), we may mention the inverse d.f. of the normal, the logistic or the double-exponential law. More generally let g be a score function of the form $g = F^{-1}$, where F is a d.f. which is subject to the following conditions

a) F is strictly increasing on $F^{-1}(]0,1[)$

b) $f := F'$ and f' exist and are assumed to be continuous and non-vanishing throughout $F^{-1}(]0,1[)$.

Let $\eta > 0$. Then the following statements (A) and (B) are equivalent:

(A) $\exists\, K > 0 \quad \forall\, t \in]0,1[: \qquad |g''(t)| \leqq K(t(1-t))^{-\eta}$

(B) $\exists K > 0 \quad \exists x_1 \leq x_2 \in \mathbb{R}^1, \quad 0 < F(x_1) \leq F(x_2) < 1$:

$$F(x) \leq K\left(\frac{f^3(x)}{|f'(x)|}\right)\eta^{-1} \qquad \forall x \leq x_1,$$

$$1 - F(x) \leq K\left(\frac{f^3(x)}{|f'(x)|}\right)\eta^{-1} \; \forall x \geq x_2 .$$

We are going to consider the class of correlation rank statistics (1.1), which is now written in the form

$$T_n := T_{n\, g_1 g_2} := \iint g_1(F_n^*) g_2(G_n^*) dH_n. \tag{2.5}$$

The main result (Theorem 2.4) of this section is preceded by two lemmas, the first of which extends the Bahadur lemma as given in Sen and Ghosh (1971). A somewhat related result attributed to van Zwet may be found in Ruymgaart (1974).

Lemma 2.1. Let $\mathcal{J}$ denote the set of all rectangles $J := J(x,y;u,v)\mathcal{H} :=]x,u[\times]y,v[\subset \mathbb{R}^2$. Then for all $\alpha \geq 1$, $\beta \in]0,1[$, $C > 0$ and $\kappa > 0$ there exists some $K > 0$ such that for sufficiently large n and for all $H \in \mathcal{H}$ it holds true that

$$P_H\left(\sup_{J \in \mathcal{J}} |H_n(J) - H(J)| \leq Kn^{-\frac{1}{2}-\frac{\beta}{2}} (\log n)^{\frac{\alpha+1}{2}} \right) \geq 1 - n^{-\kappa}. \tag{2.6}$$

$$H(J) \leq C n^{-\beta} (\log n)^{\alpha}$$

Proof: For $n \in \mathbb{N}$ let $x_{ni} := F^{-1}(i/n), y_{ni} := G^{-1}(i/n)$, $i=0,\ldots,n$ and $\mathcal{J}_n$ the set of corresponding rectangles $J_n(i,j;\, k,\ell) := J(x_{ni}, y_{nj};\, x_{nk}, y_{n\ell})$. Then for all $J := J(x,y;\, u,v) \in \mathcal{J}$ there exist integers $0 \leq i,j,k,\ell \leq n$ such that $(x,y) \in J_n(i-1,j-1;\, i,j)$, $(u,v) \in J_n(k-1,\ell-1;\, k,\ell)$. Because of

$$(H_n-H)(J_n(i,j;\, k-1,\ell-1)) - H(J_n(i-1,j-1;\, k,\ell)) + H(J_n(i,j;\, k-1,\ell-1))$$

$$\leq (H_n-H)(J) \leq (H_n-H)(J_n(i-1,j-1;\, k,\ell)) + H(J_n(i-1,j-1;\, k,\ell))$$

$$- H(J_n(i,j;\, k-1,1-\ell))$$

and

$$H(J_n(i-1,j-1;\ k,\ell)) - H(J_n(i,j;\ k-1,\ell-1))$$

$$\leqq F(]x_{n\ i-1},x_{ni}]) + F(]x_{n\ k-1},x_{nk}]) + G(]y_{n\ j-1},y_{nj}])$$

$$+ G(]y_{n\ \ell-1},y_{n\ell}]) \leqq \frac{4}{n},$$

and because of $\beta \leqq 1$ we arrive at

$$\sup_{J \in \mathcal{J}} |H_n(J)-H(J)| \leqq \max_{J_n \in \mathcal{J}_n} |H_n(J_n) - H(J_n)| + \frac{4}{n}, \tag{2.7}$$

$$H(J) \lesseqqgtr Cn^{-\beta}(\log n)^{\alpha} \qquad H(J_n) \lesseqqgtr C'n^{-\beta}(\log n)^{\alpha}$$

where we put $C' := C + 4$. Making use of $\alpha \geqq 1$, β $]0,1]$ we obtain from Bernsteins inequality, cf. Hoeffding (1963), that for each $J_n \varepsilon \mathcal{J}_n$ with $H(J_n) \leqq C'n^{-\beta}(\log n)^{\alpha}$ it holds true that

$$P_H(|H_n(J_n) - H(J_n)| > Kn^{-\frac{1}{2}-\frac{\beta}{2}}(\log n)^{\frac{\alpha+1}{2}})$$

$$\leqq 2\exp\left[\frac{-n(Kn^{-\frac{1}{2}-\frac{\beta}{2}}(\log n)^{\frac{\alpha+1}{2}})^2}{2C'n^{-\beta}(\log n)^{\alpha} + \frac{2}{3}Kn^{-\frac{1}{2}-\frac{\beta}{2}}(\log n)^{\frac{\alpha+1}{2}}}\right]$$

$$\leqq 2\exp\left[\frac{-K^2}{2(C'+K)}\log n\right] = n^{-\kappa-4}$$

for sufficiently large n and a suitably chosen K. Therefore from (2.7) and using $\alpha \geqq 1$, $\beta \in]0,1]$ we obtain for sufficiently large n

$$P_H\left(\sup_{J \in \mathcal{J}} |H_n(J)-H(J)| > Kn^{-\frac{1}{2}-\frac{\beta}{2}}(\log n)^{\frac{\alpha+1}{2}}\right) \leq$$

$H(J) \leq Cn^{-\beta}(\log n)^{\alpha}$

$$\leq P_H\left(\max_{J_n \in \mathcal{J}_n} |H_n(J_n) - H(J_n)| > Kn^{-\frac{1}{2}-\frac{\beta}{2}}(\log n)^{\frac{\alpha+1}{2}} - \frac{4}{n}\right) \leq$$

$H(J_n) \leq C'n^{-\beta}(\log n)^{\alpha}$

$$\leq \sum_{J_n \in \mathcal{J}_n} P_H\left(|H_n(J_n) - H(J_n)| > Kn^{-\frac{1}{2}-\frac{\beta}{2}}(\log n)^{\frac{\alpha+1}{2}}\right)$$

$$\leq \left(\frac{n(n+1)}{2}\right)^2 n^{-\kappa-4} \leq n^{-\kappa}.$$

<u>Remark 2.2</u> a) Along the same lines an m-dimensional analogon can be proved, $m \in \mathbb{N}$. Moreover, the assumption that the r.v. are identically distributed can be dropped. In this case H has to be replaced by the average of the theoretical d.f. throughout.

b) Lemma 2.1 is of some interest of its own. As an application that is beyond the scope of the present paper, let us mention Chernoff's estimator of the m-dimensional mode. If in Wegman's (1971) argument the law of the iterated logarithm for the discrepancy is replaced by our Lemma 2.1, we arrive at a somewhat better result than Wegman (1971) or Rüschendorf (1977).

All the technical tools that are needed in the main theorem to follow are gathered in

<u>Lemma 2.3</u> Let $\kappa > 0$, $\delta \in]0,1[\underline{x}_n := F^{-1}(n^{-1+\delta})$, $\overline{x}_n := F^{-1}(1 - n^{-1+\delta})$ and suppose that the score functions g_1 and g_2 fulfill condition (C). Then there exists a constant $K > 0$ such that

for all sufficiently large n and all $H \in \mathcal{H}$ the following assertions hold true with probability not less than $1 - n^{-\kappa}$:

$$x \in [\underline{x}_n, \overline{x}_n] \Rightarrow |F_n(x) - F(x)| \leq Kn^{-\frac{1}{2}} \log n \; r^{\frac{1}{2}}(F(x)), \tag{2.8}$$

$$x \in \mathbb{R}^1 \Rightarrow |F_n(x)-F(x)| \leq Kn^{-\frac{1}{2}} \log n \; r^{\zeta}(F(x)), \; \zeta := \frac{1}{2(\kappa+1)}, \tag{2.9}$$

$$F_n(\underline{x}_n) \in [\tfrac{1}{2} n^{-1+\delta}, \tfrac{3}{2} n^{-1+\delta}], \qquad F_n(\underline{x}_n) \leq \tfrac{1}{2}, \tag{2.10}$$

$$x \in]\underline{x}_n, \overline{x}_n] \Rightarrow r(F_n(x)) \in [\tfrac{1}{2} r(F(x)), \tfrac{3}{2} r(F(x))], \tag{2.11}$$

$$x \in]\underline{x}_n, \overline{x}_n] \Rightarrow |g_1(F_n(x))-g_1(F(x))| \leq K|F_n(x)-F(x)| r^{-1}(F(x)), \tag{2.12}$$

$$\begin{aligned} x \in]\underline{x}_n, \overline{x}_n] \Rightarrow \; & |g_1(F_n(x))-g_1(F(x))-(F_n(x)-F(x))g_1'(F(x))| \\ & \leq K(F_n(x)-F(x))^2 \, r^{-2}(F(x)), \end{aligned} \tag{2.13}$$

$$\begin{aligned} F(]x,x'[) \leq n^{-\frac{1}{2}+\frac{\delta}{4}} \log n \Rightarrow \; & |F_n(]x,x']) - F(]x,x'])| \\ & \leq Kn^{-\frac{3}{4}+\frac{\delta}{4}} \log n. \end{aligned} \tag{2.14}$$

These assertions remain true, if F_n is replaced by F_n^* and if $F, F_n, F_n^*, x, \underline{x}_n, \overline{x}_n$ and g_1 are replaced by $G, G_n, G_n^*, y, \underline{y}_n, \overline{y}_n$ and g_2.

$$H(J) \leq n^{-1+\frac{\delta}{2}} \log n \Rightarrow |H_n(J)-H(J)| \tag{2.15}$$

<u>Proof</u>: (2.8) and (2.9) are due to Ghosh (1972).

By (2.8) and the definition of $\underline{x}_n$ it follows for sufficiently large n

$$F_n(\underline{x}_n) = F(\underline{x}_n) + (F_n(\underline{x}_n)-F(\underline{x}_n)) \leqq F(\underline{x}_n)[1 + n^{\frac{1}{2}-\frac{\delta}{2}} Kn^{-\frac{1}{2}} \log n]$$

$$\leqq \frac{3}{2} n^{-1+\delta},$$

$$F_n(\underline{x}_n) \geqq F(\underline{x}_n) - |F_n(\underline{x}_n)-F(\underline{x}_n)| \geqq F(\underline{x}_n)(1 - Kn^{-\frac{\delta}{2}} \log n) \tag{2.16}$$

$$\geqq \frac{1}{2} n^{-1+\delta}.$$

Using (2.8) and $n^{-1+\delta} \leqq F(x) \leqq 1 - n^{-1+\delta}$ it follows by elementary calculations

$$r(F_n(x)) \leqq r(F(x))+2|F_n(x)-F(x)| \leqq r(F(x))(1+2r^{-\frac{1}{2}}(F(x))Kn^{-\frac{1}{2}} \log n)$$

$$\leqq r(F(x))(1 + 2n^{\frac{1}{2}-\frac{\delta}{2}}(1 - n^{-1+\delta})^{-\frac{1}{2}} Kn^{-\frac{1}{2}} \log n) \leqq \frac{3}{2} r(F(x)),$$

and $r(F_n(x)) \geqq \frac{1}{2} r(F(x))$, analogously.

(2.12) and (2.13) are easily deduced from Taylor's formula, condition (C) and $F_n(x) \in [\frac{1}{2} F(x), \frac{3}{2} F(x)]$.

(2.14) is a consequence of remark 2.2, if we put $m = 1$, $\beta = \frac{1}{2} - \frac{\delta}{2}$, $\alpha = 1$ and $C = 1$.

Take $\beta = 1 - \frac{\delta}{2}$, $\alpha = 1$, $C = 1$ and apply lemma 2.1 in order to get (2.15).

Under every fixed distribution P_H the rank statistic T_n is compared with the average of i.i.d. r.v. As usual we define

$$S_n := S_{nH} := S_{nH}^{(1)} + S_{nH}^{(2)} + S_{nH}^{(3)}, \tag{2.17}$$

where

$$S_{nH}^{(1)}: = \iint g_1(F)g_2(G)dH_n = \frac{1}{n}\sum_{i=1}^{n} g_1(F(X_i))g_2(G(Y_i)),$$

$$S_{nH}^{(2)}: = \iint (F_n-F)g_1'(F)g_2(G)dH = \frac{1}{n}\sum_{i=1}^{n}\iint (I_{\{X_i\leq x\}} - F(x))g_1'(F(x))g_2(G(y))dH(x,y)$$

$$S_{nH}^{(3)}: = \iint (G_n-G)g_2'(G)g_1(F)dH = \frac{1}{n}\sum_{i=1}^{n}\iint (I_{\{Y_i\leq y\}} - G(y))g_2'(G(y))g_1(F(x))dH(x,y).$$

We are now ready to formulate our main result.

Theorem 2.4 Let (X_1,Y_1), $(X_2,Y_2),\ldots$ be i.i.d. r.v. with d.f. $H \in \mathcal{H}$ The rank statistic T_n as defined in (2.5) is supposed to be based on score functions g_1, g_2, which satisfy condition (C). Then for all $\kappa > 0$ there exists some $\gamma > 0$ such that for all sufficiently large n one has

$$\sup_{H \in \mathcal{H}} P_H(|T_n - S_{nH}| \geqq n^{-\frac{1}{2}-\gamma}) \leqq n^{-\kappa}. \tag{2.18}$$

In particular for all $H \in \mathcal{H}$ it holds true that (as $n \to \infty$)

$$T_n - S_{nH} = O(n^{-\frac{1}{2}-\gamma}) \quad P_H\text{ - a.s.,} \qquad \exists\, \gamma > 0.$$

For the proof to follow the difference $Z_n: = Z_{nH}: = T_n - S_{nH}$ is written as $Z_n = \sum_{i=1}^{11} Z_n^{(i)}$, where

$$Z_n^{(1)} := \iint_{J_n} (g_1(F_n^*) - g_1(F) - (F_n^*-F)g_1'(F))g_2(G_n^*)dH_n$$

$$Z_n^{(2)} := \iint_{J_n} (g_2(G_n^*) - g_2(G) - (G_n^*-G)g_2'(G))g_1(F)dH_n$$

$$Z_n^{(3)} := \iint_{J_n} (F_n^*-F)g_1'(F)(g_2(G_n^*) - g_2(G))dH_n$$

$$Z_n^{(4)} := \iint_{J_n} (F_n-F)g_1'(F)g_2(G)d(H_n-H)$$

$$Z_n^{(5)} := \iint_{J_n} (G_n-G)g_2'(G)g_1(F)d(H_n-H)$$

$$Z_n^{(6)} := -\frac{1}{n+1} \iint_{J_n} F_n g_1'(F)g_2(G)dH_n$$

$$Z_n^{(7)} := -\frac{1}{n+1} \iint_{J_n} G_n g_2'(G)g_1(F)dH_n$$

$$Z_n^{(8)} := -\iint_{J_n^c} (F_n-F)g_1'(F)g_2(G)dH$$

$$Z_n^{(9)} := -\iint_{J_n^c} (G_n-G)g_2'(G)g_1(F)dH$$

$$Z_n^{(10)} := \iint_{J_n^c} g_1(F_n^*)g_2(G_n^*)dH_n$$

$$Z_n^{(11)} := -\iint_{J_n^c} g_1(F)g_2(G)dH_n$$

with $J_n :=]\underline{x}_n, \overline{x}_n] \times [\underline{y}_n, \overline{y}_n]$; $\underline{x}_n := F^{-1}(n^{-1+\delta})$, $\overline{x}_n := F^{-1}(1-n^{-1+\delta})$, $\underline{y}_n := G^{-1}(n^{-1+\delta})$, $\overline{y}_n := G^{-1}(1-n^{-1+\delta})$, where now $\delta \in]0, \frac{1}{2}[$ and arbitrary otherwise. Lemma 2.3 holds for every $\kappa > 0$. Therefore (2.18) is proved, if for each $i = 1,\dots,11$ there exists $\gamma_i > 0$ such that for sufficiently large n it holds uniformly in $H \in \mathcal{H}$

$$P_H(|Z_n^{(i)}| \geqq n^{-\frac{1}{2}-\gamma_i}) \leqq n^{-\kappa}. \qquad (2.20)$$

The most difficult terms to handle are the analogous expressions $Z_n^{(4)}$ and $Z_n^{(5)}$. The treatment of these two terms is sketched in advance: For $n \in \mathbb{N}$, $k_n := \{n^{(1/2)-(\delta/4)}\}$ and $H \in \mathcal{H}$ fixed, partition J_n into k_n^2 rectangles J_{nij} with prescribed equal marginal probabilities $k_n^{-1}(1 - 2n^{-1+\delta})$ and correspondingly write $Z_n^{(4)}$ as a sum of the integrals

$$Z_{nij}^{(4)} := \int\int_{J_{nij}} (F_n - F)g_1'(F)g_2(G)d(H_n - H), \quad i,j=1,\ldots,k_n. \qquad (2.21)$$

This partition is used to approximate the integrand of $Z_{nij}^{(4)}$ by the value it takes at a suitable lattice point belonging to J_{nij} and to rewrite the difference as an appropriate sum. In order to get upper bounds for these summands by means of condition (C) and lemma 2.3 we need upper bounds for the unknown quantities $|H_n(J_{nij})-H(J_{nij})|$ and $H_n(J_{nij})+H(J_{nij}) \leqq |H_n(J_{nij})-H(J_{nij})|+2H(J_{nij})$. For this we apply the Bahadur Lemma (2.15) realizing that J_{nij} may be exhausted by $\{q_{nij}\}$ cells of H-probability not larger than $k_n^{-2}(1-2n^{-1+\delta})^2 \leqq n^{-1+(\delta/2)}$:

$$q_{nij}: \quad \frac{H(J_{nij})}{F^{\otimes} G(J_{nij})} = H(J_{nij})k_n^2(1 - 2n^{-1+\delta})^{-2} . \qquad (2.22)$$

These quantities q_{nij} will appear as weights in the bounds we shall derive. Since they are unknown, their least favorable configuration has to be determined. Due to the fact that under the fixed marginals $F(]x_{n\,i-1}, x_{ni}]) = G(]y_{n\,j-1}, y_{nj}]) = k_n^{-1}(1 - 2n^{-1+\delta})$ only certain cell probabilities $H(J_{nij})$ can arise, we come to a linear programming problem for the appropriate standardized weights

$$p_{nij} := q_{nij}\, k_n^{-1}(1 - 2n^{-1+\delta}). \tag{2.23}$$

The solution proves to be $p_{nij} = \delta_{ij}$ (Kronecker symbol), which reduces double sums to usual sums. These turn out to be of the order $n^{-\frac{1}{2}-\gamma'}$, $\gamma' > 0$.

Proof: For sufficiently large n the assertions provided by lemma 2.3 are simultaneously available on a set of probability not less than $1 - n^{-\kappa}$. Throughout the proof we repeatedly make use of bounds like

$$\int_{]\underline{x}_n,\overline{x}_n]} r^{-2}(F)\,dF_n \leqq \left(\frac{2}{3}\right)^{-2} \int_{]\underline{x}_n,\overline{x}_n]} r^{-2}(F_n)\,dF_n$$

$$= \left(\frac{3}{2}\right)^2 \frac{1}{n} \sum_{i=\{nF_n(\underline{x}_n)\}}^{[nF_n(\overline{x}_n)]} \left(\frac{i}{n}\left(1 - \frac{i}{n}\right)\right)^{-2}$$

according to (2.11); moreover we often get estimates for sums by interpreting them as lower Riemann integrals etc. For the sake of brevity a good part of such details is skipped. In this respect, only $Z_n^{(4)}$ is treated in some more detail.

$Z_n^{(1)}$ (and analogously $Z_n^{(2)}$) can be treated by means of (2.13), (2.8), (2.1), (2.3) with $\eta \in]0, 1/2[$ and Cauchy-Schwarz's inequality as follows

$$|Z_n^{(1)}| \leqq Kn^{-1}(\log n)^2 \iint_{J_n} r^{-1}(F)r^{-\eta}(G)\,dH_n$$

$$\leqq Kn^{-1}(\log n)^2\left(\int_{]\underline{x}_n,x_n]} r^{-2}(F)\,dF_n \int_{]\underline{y}_n,\overline{y}_n]} r^{-2\eta}(G)\,dG_n\right)^{\frac{1}{2}}$$

$$\leqq Kn^{-\frac{1}{2}-\frac{\delta}{2}}(\log n)^2,$$

since $\int r^{-2\eta}(G)dG_n < \infty$ and $\int_{]\underline{x}_n,\overline{x}_n]} r^{-2}(F)dF_n = O(n^{+1-\delta})$.

$Z_n^{(3)}$ can be handled similarly. (2.12), (2.4), (2.8) and Cauchy-Schwarz's inequality yield $|Z_n^{(3)}| \leqq Kn^{-1}(\log n)^3$, since $\int_{]\underline{x}_n,\overline{x}_n]} r^{-1}(F)dF_n = O(\log n)$.

$Z_n^{(6)}$ (and analogously $Z_n^{(7)}$) is treated with the help of (2.3) with $\eta \in]0, 1/2[$, (2.4) and Cauchy-Schwarz's inequality, yielding

$$|Z_n^{(6)}| \leqq Kn^{-\frac{1}{2}-\frac{\delta}{2}}(\log n)^2.$$

$Z_n^{(8)}$ (and analogously $Z_n^{(9)}$) can be treated by subdividing J_n^c into eight rectangles, three of which are essentially different:

$$J_{no} :=]-\infty, \underline{x}_n] \times]-\infty, \underline{y}_n],\ J_{n1} :=]-\infty, \underline{x}_n] \times]\underline{y}_n, \overline{y}_n],$$

$$J_{n2} :=]\underline{x}_n, \overline{x}_n] \times]-\infty, \underline{y}_n].$$

Using (2.9) with $\zeta = 1/(2\kappa + 2)$, (2.4), (2.3) with $\eta = \zeta/(2\zeta + 2)$ and Hölder's inequality with $p = 1 + \zeta$ and $q = 1 + (1/\zeta)$ it follows

$$|Z_{no}^{(8)}| := |\int_{J_{no}} (F_n-F)g_1'(F)g_2(G)dH| \leqq Kn^{-\frac{1}{2}} \log n| \int_{J_{no}} r^{\zeta-1}(F)r^{-\eta}(G)dH|$$

$$\leqq Kn^{-\frac{1}{2}} \log n \left(\int_0^{n^{-1+\delta}} r^{\zeta^2-1}(s)ds\right)^{\frac{1}{\zeta+1}} \left(\int_0^{n^{-1+\delta}} r^{-\frac{1}{2}}(t)dt\right)^{\frac{\zeta}{\zeta+1}} \leqq Kn^{-\frac{1}{2}-\gamma'} \log n$$

with $\gamma' = (1 - \delta)(2\zeta^2 + \zeta)/(2\zeta + 2) > 0$, and analogously

$$|Z_{ni}^{(8)}| := |\int_{J_{ni}} (F_n-F)g_1'(F)g_2(G)dH| \leqq Kn^{-\frac{1}{2}-\gamma''} \log n, \quad i = 1,2,$$

with $\gamma'' = (1-\delta)\zeta^2/(\zeta+1)$ for i=1 and $\gamma'' = (1-\delta)/2$ for i=2, respectively, since e.g. $(\int_0^{n^{-1+\delta}} r^{\zeta^2-1}(s)ds)^{\frac{1}{\zeta+1}} = O(n^{\frac{(-1+\delta)\zeta^2}{(\zeta+1)}})$ and $\int r^{-\frac{1}{2}}(t)dt < \infty$.

$Z_n^{(10)}$ (and analogously $Z_n^{(11)}$) can be handled similarly to $Z_n^{(8)}$ (and $Z_n^{(9)}$). Let $Z_{ni}^{(10)}$ be defined as $Z_{ni}^{(8)}$, $i = 0,1,2$ because of $\delta < 1/2$ the assertion for all three integrals follows with (2.3) and Hölder's inequality and this for $i = 0$ and $i = 1$ with $p \in]1, 2(1-\delta)[$, $q: = p/(p-1)$ and $\eta < 1/q$ and for $i = 1$ with p and q interchanged. $Z_n^{(4)}$ (and analogously $Z_n^{(5)}$) will be treated as it was sketched above. Let $k_n: = \{n^{(1/2)-(\delta/4)}\}$ and $x_{ni}: = F^{-1}(s_{ni})$, $y_{ni}: = G^{-1}(s_{ni})$, where $s_{ni}: = n^{-1+\delta}+k_n^{-1}(1 - 2n^{-1+\delta})i$, $i = 0,\ldots,k_n$, i.e., subdivide J_n into k_n^2 rectangles $J_{nij}: =]x_{n\,i-1}, x_{ni}] \times]y_{n\,j-1}, y_{nj}]$ with equal marginal probabilities $F(]x_{n\,i-1}, x_{ni}]) = G(]y_{n\,j-1}, y_{nj}]) = k_n^{-1}(1 - 2n^{-1+\delta})$, $i,j = 1,\ldots,k_n$. Then decompose $Z_{nij}^{(4)}$ as follows:

$$Z_{nij1}^{(4)}: = (F_n(\tilde{x}_{ni})-F(\tilde{x}_{ni}))g_1'(F(\tilde{x}_{ni}))g_2(G(\tilde{y}_{nj}))\iint_{J_{nij}} d(H_n-H)$$

$$Z_{nij2}^{(4)}: = g_1'(F(\tilde{x}_{ni}))g_2(G(\tilde{y}_{nj}))\iint_{J_{nij}} (F_n(x)-F(x)-F_n(\tilde{x}_{ni})+F(\tilde{x}_{ni}))d(H_n-H)$$

$$Z_{nij3}^{(4)}: = (F_n(\tilde{x}_{ni})-F(\tilde{x}_{ni}))g_2(G(\tilde{y}_{nj}))\iint_{J_{nij}} (g_1'(F(x))-g_1'(F(\tilde{x}_{ni})))d(H_n-H)$$

$$Z_{nij4}^{(4)}: = g_2(G(\tilde{y}_{nj}))\iint_{J_{nij}} (F_n(x)-F(x)-F_n(\tilde{x}_{ni}) + F(\tilde{x}_{ni}))$$

$$(g_1'(F(x))-g_1'(F(\tilde{x}_{ni})))d(H_n-H)$$

$$Z_{nij5}^{(4)}: = (F_n(\tilde{x}_{ni})-F(\tilde{x}_{ni}))g_1'(F(\tilde{x}_{ni}))\iint_{J_{nij}} (g_2(G(y))-g_2(G(\tilde{y}_{nj})))d(H_n-H)$$

$$Z^{(4)}_{nij6}: = g_1'(F(\tilde{x}_{ni}))\iint_{J_{nij}} (F_n(x)-F(x)-F_n(\tilde{x}_{ni})+ F(\tilde{x}_{ni}))$$

$$(g_2(G(y))- g_2(G(\tilde{y}_{nj})))d(H_n-H)$$

$$Z^{(4)}_{nij7}: = (F_n(\tilde{x}_{ni})-F(\tilde{x}_{ni}))\iint_{J_{nij}} (g_1'(F(x))-g_1'(F(\tilde{x}_{ni})))$$

$$(g_2(G(y)) - g_2(G(\tilde{y}_{nj})))d(H_n-H)$$

$$Z^{(4)}_{nij8}: = \iint_{J_{nij}} (F_n(x)-F(x)-F_n(\tilde{x}_{ni})+F(\tilde{x}_{ni}))$$

$$(g_1'(F(x))-g_1'(F(\tilde{x}_{nk})))(g_2(G(y))-g_2(G(\tilde{y}_{nj})))d(H_n-H).$$

where $\tilde{x}_{ni}: = F^{-1}(\tilde{s}_{ni}): = x_{n\ i-1}$ or x_{ni} for $i \leqq k_n/2$ or $i > k_n/2$, respectively, and $\tilde{y}_{ni}$ is defined in the same way. By this special choice of $(\tilde{x}_{ni},\tilde{y}_{nj}) \in J_{nij}$ we get

$$\tilde{f}_{n1}(\tilde{x}_{ni})\tilde{f}_{n2}(\tilde{y}_{nj}) = \sup_{(x,y)\in J_{nij}} \tilde{f}_{n1}(x)\tilde{f}_{n2}(y) \tag{2.24}$$

for all functions $\tilde{f}_{n1}(x)\tilde{f}_{n2}(y)$ which arise according to condition (C) and lemma 2.3 as upper bounds for the integrands of $Z^{(4)}_{nij2},\dots,Z^{(4)}_{nij8}$. (For the sake of simplicity let us assume, that the factors which precede the integrals in $Z^{(4)}_{nij\ell}$ and which depend on n, $\tilde{x}_{ni}$ and $\tilde{y}_{ni}$, are already incorporated into the symbols $\tilde{f}_{n1}(\tilde{x}_{ni})$, $\tilde{f}_{n2}(\tilde{y}_{nj})$). On the other hand according to (2.22) each rectangle J_{nij} contains $\{q_{nij}\}$ subcells of H-probability at most $n^{-1+(\delta/2)}$. Therefore applying (2.15) to each of these subcells we get

$$|H_n(J_{nij}) - H(J_{nij})| \leqq \{q_{nij}\}\ n^{-1+\frac{\delta}{4}}, \tag{2.25}$$

i.e., for sufficiently large n it holds true that

$$H_n(J_{nij}) + H(J_{nij}) \leqq 3\, q_{nij}\, n^{-1+\frac{\delta}{2}}. \tag{2.26}$$

Hence we will be able to bound each of the double sums $\Sigma\Sigma Z^{(4)}_{nij2},\ldots,\Sigma\Sigma Z^{(4)}_{nij8}$ by expressions of the form

$$n^{-\frac{1}{2}-\gamma'}[\Sigma\Sigma p_{nij}\, \tilde{f}_{n1}(\tilde{x}_{ni})\tilde{f}_{n2}(\tilde{y}_{nj})k_n^{-1}] \tag{2.27}$$

where $\gamma' > 0$, p_{nij} is defined by (2.23) and $\tilde{f}_{n1}(\tilde{x}_{ni})$, $\tilde{f}_{n2}(\tilde{y}_{nj})$ are suitably chosen. Similarly $\Sigma\Sigma Z^{(4)}_{nij1}$ can be bounded by a term of the form

$$n^{-\frac{1}{2}-\gamma'}[\Sigma\Sigma p_{nij}\tilde{f}_{n1}(\tilde{x}_{ni})\tilde{f}_{n2}(\tilde{y}_{nj})k_n^{-1} + \Sigma\tilde{f}_{n1}(\tilde{x}_{ni})k_n^{-1}\Sigma\tilde{f}_{n2}(\tilde{y}_{nj})k_n^{-1}]. \tag{2.28}$$

To verify (2.27) as a bound for the terms $\Sigma\Sigma Z^{(4)}_{nij\ell}$, $2 \leqq \ell \leqq 8$, the case $\ell = 3$ is considered as a typical example. By (2.8), (2.3) with $\eta < (1/4) - (\delta/8)$, (2.4), a bound similar to the one in (2.12) and (2.26) it follows

$$\begin{aligned}
&|\Sigma\Sigma Z^{(4)}_{nij3}| \\
&\leqq Kn^{-\frac{1}{2}} \log n\ \Sigma\Sigma r^{\frac{1}{2}}(\tilde{s}_{ni})r^{-\eta}(\tilde{s}_{nj}) \iint_{J_{nij}} |g_1'(F(x))-g_1'(F(\tilde{x}_{ni}))|\,d(H_n+H) \\
&\leqq Kn^{-\frac{1}{2}} \log n\ \Sigma\Sigma r^{\frac{1}{2}}(\tilde{s}_{ni})r^{-\eta}(\tilde{s}_{nj})(2(r^{-1}(\tilde{s}_{ni}))^{\frac{1}{2}}(r^{-2}(\tilde{s}_{ni})k_n^{-1})^{\frac{1}{2}} \\
&\quad (H_n(J_{nij}) + H(J_{nij})) \qquad\qquad (2.29)\\
&\leqq Kn^{-\frac{1}{2}} \log n\ \Sigma\Sigma r^{-1}(\tilde{s}_{ni})r^{-\eta}(\tilde{s}_{nj})k_n^{-\frac{1}{2}}\, 3q_{nij}\, n^{-1+\frac{\delta}{2}} \\
&\leqq Kn^{-\frac{1}{2}-\eta\delta}[\Sigma\Sigma p_{nij}r^{-1}(\tilde{s}_{ni})r^{-\eta}(\tilde{s}_{nj})k_n^{-1}]\, n^{-\frac{1}{4}+\frac{\delta}{8}+\eta\delta} \log n,
\end{aligned}$$

which indeed is of the desired form (2.27). Analogously the assertion can be verified for $\ell = 2$ and $\ell = 4,\ldots,8$, using (2.3) with $\eta < (1/4) - (\delta/8)$, (2.4), (2.8), (2.12) and (2.14): In all these cases the double sum proves to be divergent under the least favorable configuration, but at the same time a factor like $n^{-(1/4)+(\delta/8)+\eta\delta}\log n$ occurs which will compensate for the divergence.

To verify (2.27) with $\gamma' = \delta/4$ as a bound of $\Sigma\Sigma Z^{(4)}_{nij1}$ we use (2.8), (2.4), (2.3) with $\eta < 1/2$ and (2.25) with $\{q_{nij}\} \leq q_{nij} + 1$ according to

$$\begin{aligned}
|\Sigma\Sigma Z^{(4)}_{nij1}| &\leq Kn^{-\frac{1}{2}}\log n\ \Sigma\Sigma r^{-\frac{1}{2}}(\tilde{s}_{ni})r^{-\eta}(\tilde{s}_{nj})|H_n(J_{nij})-H(J_{nij})| \\
&\leq Kn^{-\frac{1}{2}}\log n\ \Sigma\Sigma r^{-\frac{1}{2}}(\tilde{s}_{ni})r^{-\eta}(\tilde{s}_{nj})\{q_{nij}\}\, n^{-1+\frac{\delta}{4}} \\
&\leq Kn^{-\frac{1}{2}-\frac{\delta}{4}}[\Sigma\Sigma p_{nij}\, r^{-\frac{1}{2}}(\tilde{s}_{ni})r^{-\eta}(\tilde{s}_{nj})k_n^{-1} \\
&\quad + \Sigma r^{-\frac{1}{2}}(\tilde{s}_{ni})k_n^{-1}\ \Sigma r^{-\eta}(\tilde{s}_{nj})k_n^{-1}]\log n.
\end{aligned} \tag{2.30}$$

Here in contrary to (2.29) all appearing sums turn out to be convergent.

In order to prove (2.20) the convergence (or divergence at a certain rate) of the sums in (2.30) (and (2.29)) has to be shown. For this we have to take into account the least favorable configuration of the cell probabilities $H(J_{nij})$, which are unknown up to the fact that

$$\sum_{j=1}^{k_n} H(J_{nij}) \leq F(]x_{n\ i-1}, x_{ni}]) = k_n^{-1}(1 - 2n^{-1+\delta}),$$

$$\sum_{i=1}^{k_n} H(J_{nij}) \leq G(]y_{n\ j-1}, y_{nj}]) = k_n^{-1}(1 - 2n^{-1+\delta}).$$

This means that we have to consider the least favorable matrix $(p_{nij})_{i,j=1,\ldots,k_n}$, i.e. the matrix which maximizes

$$\Sigma\Sigma p_{nij}\ \tilde{f}_1(\tilde{x}_{ni})\tilde{f}_2(\tilde{y}_{nj}) \tag{2.31}$$

under the side conditions

$$p_{ni.} \leqq 1, \qquad i = 1,\ldots,k_n, \tag{2.32}$$

$$p_{n\cdot j} \leqq 1, \qquad j = 1,\ldots,k_n. \tag{2.33}$$

$$p_{nij} \geqq 0, \qquad i,j = 1,\ldots,k_n. \tag{2.34}$$

This is for each $n \in \mathbb{N}$ a linear programming problem for the double substochastic matrix $(p_{nij})_{i,j=1,\ldots,k_n}$, for which the optimal value is attained over the set of its extremal points, i.e., over the set of permutation matrices. Since we are only interested in the convergence rate of an upper bound of (2.31), we can ignore that there is also a lower bound for $p_{ni\cdot}$ in (2.32) according to the fact that

$$\sum_{j=1}^{k_n} H(J_{nij}) \geqq F(]x_{n\ i-1}, x_{ni}]) - 2n^{-1+\delta}, \quad i=1,\ldots,k_n;$$

similarly for $p_{n\cdot j}$ in (2.25)). This implies that the maximum of (2.31) under the side conditions (2.32) - (2.34) equals

$$\max \sum_{i=1}^{k_n} \tilde{f}_{n1}(\tilde{x}_{ni})\tilde{f}_{n2}(\tilde{y}_{nj(i)}), \tag{2.35}$$

where this maximum is taken over the set of all permutations $(j(1),\ldots,j(k_n))$ of the integers $(1,\ldots,k_n)$. But this becomes

$$\sum_{i=1}^{k_n} \tilde{f}_{n1}(\tilde{x}_{ni})\tilde{f}_{n2}(\tilde{y}_{ni}), \tag{2.36}$$

since $(\tilde{f}_{n1}(\tilde{x}_{n1}),\dots,\tilde{f}_{n1}(\tilde{x}_{nk_n}))$ and $(\tilde{f}_{n2}(\tilde{y}_{n1}),\dots,\tilde{f}_{n2}(\tilde{y}_{nk_n}))$ are ordered in the same way.

Therefore (2.27) - up to the factor $Kn^{-\frac{1}{2}-\gamma'}$ - can be bounded by

$$\sum_{i=1}^{k_n} \tilde{f}_{n1}(\tilde{x}_{ni})\tilde{f}_{n2}(\tilde{y}_{ni})k_n^{-1} \leqq 2 \sum_{i=2}^{[\frac{k_n}{2}]} \tilde{f}_{n1}(\tilde{x}_{ni})\tilde{f}_{n2}(\tilde{y}_{ni})k_n^{-1} + 2\tilde{f}_{n1}(\tilde{x}_{n1})\tilde{f}_{n2}(\tilde{y}_{n1})k_n^{-1} . \tag{2.37}$$

Since $\tilde{f}_{ni}(x)$ is non-increasing for $x \leqq 1/2$, $i = 1,2$, $n \in \mathbb{N}$, the first summand on the right side of (2.37) can be bounded for each $n \in \mathbb{N}$ by $2 \int_{]\underline{x}_n,F^{-1}(\frac{1}{2})]} \tilde{f}_{n1}(x)\tilde{f}_{n2}(x)dF(x)$, whereas the second can be evaluated directly. According to (2.29) in the case $\ell = 3$ we have $\tilde{f}_{n1}(x)\tilde{f}_{n2}(x) = r^{-1-\eta}(F(x))\, n^{-\frac{1}{4}+\frac{\delta}{8}+\eta\delta} \log n$, i.e., the upper bound contains the divergent integral $2\int_{n^{-1+\delta}}^{1/2} r^{-1-\eta}(s)ds = O(n^{\eta-\eta\delta})$, which is of smaller order of divergence than the convergence to 0 of the factor $n^{-\frac{1}{4}+\frac{\delta}{8}+\eta\delta} \log n$ for $\eta < 3/16$. Correspondingly the second term on the right side of (2.37) tends to zero.

Analogously, (2.30) - up to the factor $Kn^{-\frac{1}{2}-\frac{\delta}{4}}\log n$ - can be bounded by $2\int_{n^{-1+\delta}}^{1/2} r^{-\frac{1}{2}-\eta}(s)ds + 4\int_{n^{-1+\delta}}^{1/2} r^{-\frac{1}{2}}(s)ds \int_{n^{-1+\delta}}^{1/2} r^{-\eta}(s)ds$, which is convergent for $n \to \infty$, and summands corresponding to the outer cells in (2.28), which tend to zero for $n \to \infty$.

Remark 2.5 So far, only the approximative scores $g_i(\frac{j}{n+1})$ were utilized. However, by a reasoning similar to the one by Chernoff-Savage (1958), p. 991-994, everything will remain valid if T_n is based on the exact scores $E(g_i(U_{n[j]}))$ instead.

In consequence of the foregoing theorem a number of statements on the asymptotic behavior of T_n can be made. We are going to single out just a few of them. First, let us put

$$\mu(H) := \iint g_1(F)g_2(G)dH, \quad \sigma^2(H) := \mathrm{var}_H(S_{1H}) \tag{2.38}$$

whenever these quantities are defined. Note that by condition (C)

$$\mu(H) = E_H(S_{1H}), \quad \sup_{H\in\mathcal{H}} E_H|S_{1H}|^3 < \infty. \tag{2.39}$$

Corollary 2.6 On the conditions of Theorem 2.4 and if $\sigma(H) > 0$, then

$$\overline{\lim_n} \frac{n^{1/2}(T_n-\mu(H))}{(2\log\log n)^{1/2}} = \sigma(H) \qquad TP_H\text{ - a.s.}$$

(and similarly for "$\underline{\lim}$.")

Remark 2.7 If $H(x,y) = F(x)G(y)$, a law of the iterated logarithm for $(T_n)_{n\geqq 1}$ was proved by Sen and Ghosh (1974).

Corollary 2.8 Suppose that $(N_n)_{n \geqq 1}$ is a sequence of $\mathbb{N}$-valued random indices for which N_n/d_n converges in probability to some positive r.v. τ, where $(d_n)_{n \geqq 1} \subset \mathbb{R}^1_+$, $d_n \uparrow \infty$ (as $n \to \infty$). On the conditions of Theorem 2.4,

$$L_H(N_n^{\frac{1}{2}}(T_{N_n} - \mu(H))) \underset{\mathcal{D}}{\to} N(0,\sigma^2(H)). \tag{2.40}$$

In particular, if $N_n = n$, $d_n = n$ for $n \geqq 1$,

$$L_H(n^{\frac{1}{2}}(T_n - \mu(H))) \underset{\mathcal{D}}{\to} N(0,\sigma^2(H)). \tag{2.41}$$

Evidently, none of the above corollaries can be concluded from a Chernoff-Savage theorem in the usual $o_p(n^{-1/2})$ version. This is also true for the functional limit theorem which we are going to formulate below.

Let $\{H_\Delta: 0 \leqq \Delta \leqq \Delta_o\} \subset H$ be a real parameter family of d.f. To simplify the notation we shall write $E_\Delta(S_{1\Delta})$ instead of $E_{H_\Delta}(S_{1H_\Delta})$ etc. Moreover, it is required that (as $\Delta \to 0$):

$$E_\Delta(S_{1\Delta}) = \Delta\xi + o(\Delta), \quad \exists\, \xi \in \mathbb{R}^1, \tag{2.42}$$

$$\mathrm{Var}_\Delta(S_{1\Delta}) = \sigma^2 + o(1), \quad \exists\, \sigma > 0. \tag{2.43}$$

We shall start out from the function space $C[0,\infty[$ which is supposed to be equipped with the Borel sets generated by the usual compact-open topology. Throughout, W will denote a standard Wiener process defined on this space. Introduce random functions W_Δ as follows. For all $j \geqq 0$ and for $j\Delta^2 \leqq t \leqq (j+1)\Delta^2$ let

$$W_\Delta(t) = j\Delta T_j + (t\Delta^{-2} - j)((j+1)\,\Delta\, T_{j+1} - j\Delta T_j), \quad (T_o = 0) \tag{2.44}$$

Theorem 2.9 Let us write s for the identity map restricted to $[0,\infty[$. On the conditions of Theorem 2.4 and if (2.42), (2.43) are satisfied, then as $\Delta \to 0$,

$$W_\Delta \underset{\mathcal{D}}{\to} \sigma W + \sigma\xi s \tag{2.45}$$

Proof. A preliminary version of this theorem is given in Bönner (1976), the present one follows from Müller-Funk (1979), Theorem 3.4. Alternatively, we may refer to Lai (1978) Theorem 3 and mimic the proof of his Theorem 4.

Obviously, theorem 2.4 implies the a.s. convergence of T_n towards $\mu(H)$. With a view to the sequential analysis applications to follow, however, a direct argument on less restrictive assumptions is desirable. The following theorem is restated from Sen and Ghosh (1974), but provided with a simpler proof.

Theorem 2.10 Suppose that g_1, g_2 are non-decreasing, continuous and square integrable score functions. For the sake of convenience, we assume that

$$\int_o^1 g_i(t)dt = 0, \quad \int_o^1 g_i^2(t)dt = 1 \qquad (i = 1,2). \tag{2.46}$$

Then for every $H \in \mathcal{H}$,

$$T_n \to \mu(H) \qquad P_H - \text{a.s.}$$

Proof. W.l.o.g. we may assume that g_1, g_2 are not identically zero. Hence there exists some $0 < \eta_o < \frac{1}{2}$ such that g_1, g_2 are negative in $]0, \eta_o[$ and positive in $]1-\eta_o, 1[$. Let $\varepsilon > 0$ and choose $0 < \eta = \eta(\varepsilon) < \eta_o$ such that

$$\int_0^{\eta} g_i^2(t)dt + \int_{1-\eta}^{1} g_i^2(t)dt < \varepsilon^2/64 \qquad (i = 1,2). \tag{2.47}$$

Put $I = [F^{-1}(\eta), F^{-1}(1-\eta)] \times [G^{-1}(\eta), G^{-1}(1-\eta)]$. By definition, cf. (2.5) and (2.38),

$$\mu(H) = \iint g_1(F)g_2(G)dH, \qquad \mu(H_n^*) = \frac{n}{n+1} T_n.$$

Now,

$$\begin{aligned} |\mu(H_n^*) - \mu(H)| &\leq \iint_I |g_1(F_n^*)g_2(G_n^*) - g_1(F)g_2(G)|dH_n^* \\ &+ \left|\iint_I g_1(F)g_2(G)dH_n^* - \iint_I g_1(F)g_2(G)dH\right| \\ &+ \iint_{I^c} |g_1(F_n^*)g_2(G_n^*)|dH_n^* + \iint_{I^c} |g_1(F)g_2(G)|dH \\ &= a_n + b_n + c_n + d_n. \end{aligned}$$

By the Glivenko-Cantelli-Theorem and the assumed continuity of g_1, g_2, we conclude from Polya's Theorem that the integrand of a_n uniformly approaches zero and hence $a_n < \varepsilon/4$ for n sufficiently large. By the Helly-Bray Theorem, the latter can also be ensured for b_n. Taking into account that g_1^2, g_2^2 are non-increasing in $]0,\eta[$ and non-decreasing in $]1-\eta,1[$, we infer from the Cauchy-Schwartz inequality and (2.47) that $|c_n| < \varepsilon/4$ for n sufficiently large. Similarly by (2.47), this can be shown to hold true for d_n.

III. APPLICATIONS TO SEQUENTIAL ANALYSIS

We are going to introduce some subclasses of $\mathcal{H}$ which were discussed by Lehmann (1966). Let $\mathcal{H}_p \subset \mathcal{H}$ (resp. $\mathcal{H}_n \subset \mathcal{H}$) be the set of positively (resp. negatively) quadrant dependent d.f., i.e. $H(x,y) \geqq F(x)G(y)$ (resp. $\leqq$). Here "$\geqq$" means $\geq$ for all (x,y) with strict inequality for at least one point. The family of all d.f. $H \in \mathcal{H}_p$ (resp. $H \in \mathcal{H}_n$) such that one component is positively (resp. negatively) regression dependent on the other one, will be denoted by $\mathcal{H}_+$ (resp. $\mathcal{H}_-$ Finally, the class of all $H \in \mathcal{H}$ which corresponds to the independence case will be labelled $\mathcal{H}_u$. We shall consider the following testing problem:

$$A_o: \; H \in \mathcal{H}_n \cup \mathcal{H}_u, \qquad A_1: \; H \in \mathcal{H}_p.$$

Our testing procedures will be based on the rank statistics $T_n = T_{ng_1g_2}$. First we point at the fact that by (2.41) T_n is asymptotically normal with asymptotic shift parameter $\mu(H)$. In case that g_1, g_2 are non-decreasing, $\mu(\cdot)$ correctly reflects the stochastic ordering expressed by the hypotheses above, i.e.

$$H \in \mathcal{H}_p \quad \Rightarrow \quad \mu(H) \geqq 0,$$

$$H \in \mathcal{H}_u \quad \Rightarrow \quad \mu(H) = 0,$$

$$H \in \mathcal{H}_n \quad \Rightarrow \quad \mu(H) \leqq 0.$$

Accordingly, it is suggestive to pattern a sequential test upon the SPRT for testing a normal mean. More precisely, we are guided by the simpler testing problem

$$A_0': \ \mu(H) = 0, \qquad A_1': \ \mu(H) = \theta > 0$$

as for the structure of the sequential tests we are looking for. In doing so, we fix $b < 0 < a$, $\theta > 0$ and define for every $\underset{\sim}{t} = (t_n)_{n \geqq 1} \in \mathbb{R}^{\mathbb{N}}$, $(\inf \phi: = \infty)$:

$$\overline{N}_a(\theta,\underset{\sim}{t}): = \min\{n \geqq 1: \ n\theta(t_n - \frac{\theta}{2}) \geqq a\},$$

$$\underline{N}_b(\theta,\underset{\sim}{t}): = \min\{n \geqq 1: \ n\theta(t_n - \frac{\theta}{2}) \leqq b\},$$

$$N(\theta,\underset{\sim}{t}): \ = \min\{\overline{N}_a(\theta,\underset{\sim}{t}), \ \underline{N}_b(\theta,\underset{\sim}{t})\}.$$

If $\underset{\sim}{T} = (T_n)_{n \geqq 1}$ is inserted, we shall write $N(\theta): = N(\theta,\underset{\sim}{T})$, $\overline{N}_a(\theta): = \overline{N}_a(\theta,\underset{\sim}{T})$ and $\underline{N}_b: = \underline{N}_b(\theta,\underset{\sim}{T})$. Now, as in Sen and Ghosh (1974) those sequential tests with stopping time $N(\theta)$ are considered which accept A_o or A_1 according as $\underline{N}_b(\theta) < \overline{N}_a(\theta)$ or $\underline{N}_b(\theta) > \overline{N}_a(\theta)$. For short, such a test is termed a sequential correlation rank test (SCRT) with score functions g_1,g_2.

As in the non-sequential case, only few statements can be made which are not asymptotic in nature.

<u>Theorem 3.1</u> Suppose g_1,g_2 are non-decreasing. Put

$$\phi_n(\theta): = I_{\{n\theta(T_n - \frac{\theta}{2}) \geqq a\}}, \qquad n \in \mathbb{N}, \quad \Theta > 0.$$

a) If $\mu(H) \neq \frac{\theta}{2}$ and if T_n converges towards $\mu(H)$ P_H-a.s.,

or

if $\mu(H) = \frac{\theta}{2}$ and if T_n is asymptotically normal with parameters $(\mu(H), \sigma^2(H)/n)$, $\sigma^2(H) > 0$,

then the SCRT terminates P_H -a.s. In particular this holds true if the assumption of theorem 2.10 resp. corollary 2.8 are fulfilled.

b) For every $H_+ \in \mathcal{H}_+$, $H_o \in \mathcal{H}_u$ and $H_- \in \mathcal{H}_-$ it holds true that

$$E_{H_+}(\phi_{N(\theta)}(\theta)) \geq E_{H_o}(\phi_{N(\theta)}(\theta)) \geq E_{H_-}(\phi_{N(\theta)}(\theta)).$$

Note that under $\mathcal{H}_u$ the test statistics are distribution free and hence the tests are unbiased at the level $\alpha' := E_{H_o}(\phi_{N(\Theta)}(\Theta))$.

<u>Proof.</u> a) follows from a standard argument.
In order to prove b), we are going to extend the reasoning as given in Witting-Nölle (1970), Satz 3.14 to the sequential case. The last mentioned result as well as its proof remain true if the sample space $\mathbb{R}^{\mathbb{N}}$ instead of $\mathbb{R}^n$ is assumed. Taking sequences $\underset{\sim}{x}, \underset{\sim}{y}, \underset{\sim}{y}' \in \mathbb{R}^{\mathbb{N}}$ for which it is always true that

$$x_j < x_k, \quad y_j < y_k \Rightarrow y_j' < y_k', \tag{3.1}$$

we can infer from Lehmann (1966), corollary 2 of theorem 5, that $t_n' := T_n(x_1,\ldots,x_n, y_1',\ldots,y_n') \geq t_n := T_n(x_1,\ldots,x_n, y_1,\ldots,y_n)$. Note, that $\overline{N}_a(\Theta, \underset{\sim}{t})$ is non-increasing and $\underline{N}_b(\Theta, \underset{\sim}{t})$ is non-decreasing in $\underset{\sim}{t}$, $\underset{\sim}{t} = (t_n)_{n\geq 1}$. As the SCRT accepts A_1 iff $\overline{N}_a(\Theta) < \underline{N}_b(\Theta)$, we realize that the test function is non-decreasing if (3.1) is valid. Similarly, the other monotonicity properties required in the extended version of the above quoted Satz 3.14 can be shown to hold true.

Next we shall derive limiting expressions for the OC- and the ASN-curves under families $\{H_\Delta: 0 \leqq \Delta \leqq \Delta_o\}$ (not necessarily in $\mathcal{H}_n \cup \mathcal{H}_u \cup \mathcal{H}_p$) as $\Delta \to 0$. First, the reader is reminded of some formulas due to Dvoretzky, Kiefer and Wolfowitz (1953). With $b < 0 < a$ let us associate functions τ_a, τ_b defined on the space $C[0,\infty[$,

$$\tau_c(x) := \inf\{t > 0: x(t) = c\}, \qquad c \in \mathbb{R}^1,$$

$$\tau(x) := \min\{\tau_a(x), \tau_b(x)\}.$$

As it is well known these stopping times are continuous with respect to Wiener measure. It can be drawn from the above mentioned paper that for every $\upsilon \in \mathbb{R}^1$ (and with s defined as in theorem 2.9):

$$\Psi(\upsilon) := P(\tau_b(W+\upsilon s) < \tau_a(W+\upsilon s)) = \begin{cases} \dfrac{e^{-2\upsilon a}-1}{e^{-2\upsilon a}-e^{-2\upsilon b}}, & \upsilon \neq 0, \\[2ex] \dfrac{a}{a-b}, & \upsilon = 0. \end{cases} \tag{3.2}$$

$$\Gamma(\upsilon) := E(\tau(W+\upsilon s)) = \begin{cases} \dfrac{be^{-2\upsilon a} - ae^{-2\upsilon b} + a-b}{\upsilon(e^{-2\upsilon a} - e^{-2\upsilon b})}, & \upsilon \neq 0, \\[2ex] -ab, & \upsilon = 0. \end{cases} \tag{3.3}$$

Given error probabilities $\alpha > 0$, $\beta > 0$, $\alpha + \beta < 1$ the pair of equations $\Psi(-\frac{1}{2}) = 1 - \alpha$, $\Psi(+\frac{1}{2}) = \beta$ has the unique solution

$$a = \log \frac{1-\beta}{\alpha}, \qquad b = \log \frac{\beta}{1-\alpha}. \tag{3.4}$$

Theorem 3.2 If the assumptions of theorem 2.4 are fulfilled and if $\{H_\Delta: 0 \leq \Delta \leq \Delta_o\} \subset \mathcal{H}$ is a one-parameter family satisfying (2.42), (2.43) and with $\xi \in \mathbb{R}^1$ and $\sigma = 1$, then for $\eta \geq 0$ (fixed) it holds true that (as $\Delta \to 0$)

a) $P_{\Delta\eta}(\underline{N}_b(\Delta) < \overline{N}_a(\Delta)) \to \Psi(\eta\xi - \frac{1}{2})$

b) $E_{\Delta\eta}(\Delta^2 N(\Delta)) \to \Gamma(\eta\xi - \frac{1}{2}),$ if $\eta\xi \neq \frac{1}{2}$.

c) If (2.42) is replaced by $E_{\Delta\eta}(S_{1\Delta}) = \Delta\xi$, then

$$E_{\Delta\eta}(\Delta^2 N(\Delta)) \to \Gamma(0) = -ab, \qquad \text{if } \eta\xi = \frac{1}{2}.$$

Proof. a) is an immediate consequence of theorem 2.9 and of the simple relation $0 \leq \tau(W_\Delta) - \Delta^2 N(\Delta) \leq \Delta^2$. This entails that $\Delta^2 N(\Delta)$ converges in distribution towards the first time at which a Wiener process with shift parameter $\upsilon = \eta\xi - \frac{1}{2}$ leaves $]b,a[$. (Take into account that in (2.45) we now have $\eta\xi$ instead of ξ whereas $-1/2$ is due to the definition of the SCRT). In order to prove b), we have to verify that $\{\Delta^2 N(\Delta): 0 \leq \Delta \leq \Delta_1\}$ is uniformly integrable under $P_{\Delta\eta}$ for a suitably chosen $0 < \Delta_1 \leq \Delta_o$. Obviously, it suffices to show that for t sufficiently large

$$P_{\Delta\eta}(\Delta^2 N(\Delta) > t) \leq Kt^{-2}, \qquad \forall \; 0 \leq \Delta \leq \Delta_1. \tag{3.5}$$

The idea of proof will be somewhat similar to that of theorem 1 in Lai (1975b). Put $n_j := n_j(\Delta,t) := [\frac{jt}{4\Delta^2}]$, $1 \leq j \leq 4$, $n_o := 0$.

$$P_{\Delta\eta}(N(\Delta) > [t\Delta^{-2}]) \leq P_{\Delta\eta}(T_n - \frac{\Delta}{2}\epsilon]n^{-1}\Delta^{-1}b, n^{-1}\Delta^{-1}a[, \; \forall\, 1 \leq n \leq n_4)$$

$$\leq P_{\Delta\eta}(S_{n_i\Delta\eta} - \frac{\Delta}{2}\epsilon]n_i^{-1}\Delta^{-1}b - n_i^{-\frac{1}{2}-\gamma}, \; n_i^{-1}\Delta^{-1}a + n_i^{-\frac{1}{2}-\gamma}[, \quad \forall\, 1 \leq i \leq 4)$$

$$+ P_{\Delta\eta}(|T_n - S_{n\Delta\eta}| > n^{-\frac{1}{2}-\gamma}, \quad \exists\, n \geq n_1). \tag{3.6}$$

By means of theorem 2.4, it is easily seen that the second summand is of the correct order of magnitude. Now, in case $\eta\xi > \frac{1}{2}$ the first summand is treated as follows:

$$P_{\Delta\eta}(\Delta^{-1}b - n_i^{\frac{1}{2}-\gamma} + n_i(\frac{\Delta}{2} - \mu_{\Delta\eta}) < (n_i S_{n_i\Delta\eta} - n_i\mu_{\Delta\eta})$$

$$< \Delta^{-1}a + n_i^{\frac{1}{2}-\gamma} + n_i(\frac{\Delta}{2} - \mu_{\Delta\eta}), \quad \forall\, 1 \leq i \leq 4)$$

$$\leq \prod_{i=1}^{4} P_{\Delta\eta}\left(\frac{n_i S_{n_i\Delta\eta} - n_{i-1}S_{n_{i-1}\Delta\eta} - \mu_{\Delta\eta}(n_i - n_{i-1})}{(n_i - n_{i-1})^{\frac{1}{2}}\sigma_{\Delta\eta}} \leq c_i \right)$$

where
$$c_i = \frac{\Delta^{-1}(a-b) + n_i^{\frac{1}{2}-\gamma} + n_{i-1}^{\frac{1}{2}-\gamma} + (n_i - n_{i-1})(\frac{\Delta}{2} - \mu_{\Delta\eta})}{(n_i - n_{i-1})^{\frac{1}{2}}\sigma_{\Delta\eta}}$$

$$\leq \frac{\Delta^{-1}(a-b) + 2n_i^{\frac{1}{2}-\gamma} + (n_i - n_{i-1})\Delta(\frac{1}{2} - \eta\xi + o(1))}{(n_i - n_{i-1})^{\frac{1}{2}}(1 + o(1))}$$

$$\leq K_1 - t^{\frac{1}{2}} K_2, \qquad 1 \leq i \leq 4,$$

for some positive constants K_1, K_2 whenever $\Delta < \Delta_1$, $t > t_1$. Hence by the Berry-Esseen bound

$$\prod_{1}^{4} P_{\Delta\eta}((n_i - n_{i-1})^{\frac{1}{2}} \sigma_{\Delta\eta}^{-1}(S_{n_i - n_{i-1},\Delta\eta} - \mu_{\Delta\eta}) \leqq c_i)$$

$$\leqq \prod_{1}^{4} (\phi(K_1 - K_2 t^{\frac{1}{2}}) + K_3(n_i - n_{i-1})^{-\frac{1}{2}}) \tag{3.7}$$

$$\leqq (K_4 t^{-1} + K_5 t^{-\frac{1}{2}})^4 \leqq K_6 t^{-2}$$

for suitably chosen positive constants $K_1,\ldots,K_6$ and whenever $\Delta < \Delta_1$, $t > t_1$. (3.6), (3.7) together yield (3.5), whence the result follows. The case $\eta\xi < \frac{1}{2}$ can be treated analogously. If $\eta\xi = \frac{1}{2}$, this is essentially Lai's (1975b) result.

Remark 3.3 a) The same method of proof with $m := [2k+1] + 1$ instead of $m = 4$ yields that $\{\Delta^{2k} N^k(\Delta): 0 \leqq \Delta \leqq \Delta_o\}$ is in fact uniformly integrable for all $k \geqq 1$.

b) A general result on the limiting behavior of stopping times of SPRT-type tests is also contained in Lai (1978).

So far, we have not yet looked for a statistical model to which the theorem above applies, i.e., for a family of distributions in H_p (resp. H_n) which fulfills condition (2.42), (2.43). In the non-sequential case, a sufficiently broad class of non-parametric alternatives were defined by Behnen (1972). We are going to adapt his device to the present situation. Let h_i $(i = 1,2)$ be score functions satisfying (2.46) as well as

$$|h_i(t)| \leqq K(t(1-t))^{-\frac{1}{4}+\delta} \qquad \forall\ 0 < t < 1,\ \exists\, \delta > 0, \tag{3.8}$$

$$\int_0^s h_i(t)dt \leqq 0 \qquad \forall\ 0 < s < 1 \qquad (i = 1,2). \tag{3.9}$$

Put

$$h^{o}_{i\Delta}(t) = h_i(t)1_{]\Delta^{1/2},\ 1-\Delta^{1/2}[}(t),$$
$$h_{i\Delta}(t) = c_i(\Delta)[h^{o}_{i\Delta}(t) - \int_0^1 h^{o}_{i\Delta}(s)ds, \tag{3.10}$$

$0 < \underline{c} \leq c_i(\Delta) \leq \overline{c} < \infty$, where $c_i(\Delta)$ is to be determined later on. Define d.f. H_Δ

$$dH_\Delta := (1 + \Delta h_{1\Delta}(F)h_{2\Delta}(G))d(F \otimes G), \quad 0 \leq \Delta \leq \Delta_o, \tag{3.11}$$

where F,G are two continuous univariate d.f.

By (3.9) we make sure that H_Δ is actually in $\mathcal{H}_p$. Let us mention that we could have defined d.f. $H_\Delta \in \mathcal{H}_n$ for $\Delta < 0$ as well.

It should be stressed, that in fact (h_1,h_2) but not the approximating $(h_{1\Delta},h_{2\Delta})$ is the quantity we are primarily interested in. Intuitively, (h_1,h_2) describes the direction in which the alternatives depart from the hypothesis. $(h_{1\Delta},h_{2\Delta})$ is solely introduced for technical reasons and all that matters is that $\{(h_{1\Delta},h_{2\Delta}): 0 \leq \Delta \leq \Delta_o\}$ shows the correct asymptotic behaviour as $\Delta \to 0$. The asymptotic performance of the SCRT with score functions g_1,g_2 is studied under (h_1,h_2) - alternatives (3.11).

<u>Corollary 3.4.</u> Let g_1,g_2 be score functions which fulfill condition (C) as well as (2.46). Given h_1,h_2 satisfying (2.46), (3.8), (3.9) define alternatives $\{H_\Delta: 0 \leq \Delta \leq \Delta_o\}$ as in (3.11) where now $c_i(\Delta)$ are determined by the requirement that

$$\int_o^1 g_i(t)h_{i\Delta}(t)dt \equiv \int_o^1 g_i(t)h_i(t)dt =: \rho_i, \quad i = 1,2.$$

Then theorem 3.2 is in force with $\xi = \rho_1\rho_2$. In case error probabilities $\alpha > 0$, $\beta > 0$, $\alpha + \beta < 1$ are given and we choose $a := \log((1-\beta)/\alpha)$ and $b := \log(\beta/(1-\alpha))$, then by (3.4) (as $\Delta \to 0$)

$$P_0(\overline{N}_a(\Delta) < \underline{N}_b(\Delta)) \to \alpha, \qquad P_\Delta(\underline{N}_b(\Delta) < \overline{N}_a(\Delta)) \to \beta. \tag{3.12}$$

The SCRT was introduced on purely intuitive grounds but up to now it lacks a proper justification. Starting out from a class $\{H_\Delta : 0 \leq \Delta \leq \Delta_o\}$ of alternatives defined as in (3.11), the Wald SPRT is based on the statistics

$$L_{n\Delta} = \prod_{i=1}^{n} (1 + \Delta h_{1\Delta}(F(X_i))h_{2\Delta}(G(Y_i))); \quad n \geq 1. \tag{3.13}$$

By elementary but tedious claculations we get (as $\Delta \to 0$):

$$E_{\Delta\eta}(\log L_{1\Delta}) = \Delta^2(\eta - \tfrac{1}{2})(1 + O(\Delta^{1/4})),$$

$$\mathrm{Var}_{\Delta\eta}(\log L_{1\Delta}) = \Delta^2(1 + O(\Delta^{1/4})),$$

$$E_{\Delta\eta}(|\log L_{1\Delta}|^3) = o(1).$$

The stopping variables associated with the SPRT based on $(L_{n\Delta})_{n\geq 1}$ are termed $\overline{N}^*_a(\Delta)$, $\underline{N}^*_b(\Delta)$, $N^*(\Delta)$. Note that the SPRT terminates a.s.

<u>Theorem 3.5</u> Assume alternatives $\{H_\Delta : 0 \leq \Delta \leq \Delta_o\}$ as defined in (3.10), where h_1, h_2 are supposed to satisfy (3.8) - (3.10) and $c_i(\Delta)$ are determined by $\int_0^1 h_i(t)h_{i\Delta}(t)\,dt \equiv 1$. Then for $\eta \geq 0$ (fixed) it holds true that (as $\Delta \to 0$)

a) $P_{\Delta\eta}(\underline{N}^*_b(\Delta) < \overline{N}^*_a(\Delta)) \to \Psi(\eta - \frac{1}{2})$

b) $E_{\Delta\eta}(\Delta^2 N^*(\Delta)) \to \Gamma(\eta - \frac{1}{2})$ if $\eta \neq \frac{1}{2}$.

Proof. Essentially, the steps leading to theorem 3.2 are repeated.

Let $\underset{\sim}{V} = (V_n)_{n \geq 1}$, $V_n = V_n(X_1, Y_1, \ldots, X_n, Y_n)$, be a sequence of real-valued statistics. Put $\overset{\vee}{\overline{N}}_a(\theta) := \overline{N}_a(\theta, \underset{\sim}{V})$ etc. By the Wald test (cf. Lai (1978)) associated with $\underset{\sim}{V}$ and b, a, θ, we shall understand the sequential test with stopping rule $\overset{\vee}{N}(\theta)$ which decides in favour of A_o (resp. A_1) iff $\overset{\vee}{\underline{N}}_b(\theta) < \overset{\vee}{\underline{N}}_a(\theta)$ (resp. $\overset{\vee}{\overline{N}}_a(\theta) < \overset{\vee}{\underline{N}}_b(\theta)$).

Given a pair of score functions h_1, h_2 satisfying (2.46), (3.8), (3.9) we shall denote by $\mathcal{W}_{h_1,h_2}$ the class of those Wald tests such that under (h_1,h_2) - alternatives (3.11) a) the tests terminate $P_{\Delta\eta}$ -a.s. and b) the following limit expressions are valid with some $\zeta \in \mathbb{R}^1$ and $\eta \geq 0$:

$$P_{\Delta\eta}(\underline{N}_b(\Delta) < \overline{N}_a(\Delta)) \to \Psi(\eta\zeta - \frac{1}{2}),$$

$$E_{\Delta\eta}(\Delta^2 N(\Delta)) \to \Gamma(\eta\zeta - \frac{1}{2}), \text{ if } \eta\zeta \neq \frac{1}{2}.$$

Note that on the conditions of theorem 2.4 every SCRT with score functions g_1, g_2 is in $\mathcal{W}_{h_1 h_2}$ and that the SPRT (3.13) for testing $\eta = 0$ against $\eta = 1$ is in $\mathcal{W}_{h_1,h_2}$.

Corollary 3.6 If h_1, h_2 satisfy condition (C) as well as (2.46), (3.8), (3.9), then the SCRT with score functions h_1, h_2 is asymptotically optimal against (h_1,h_2)-alternatives (3.11) among all sequential tests within the class $\mathcal{W}_{h_1,h_2}$.

To complete the discussion of the sequential tests proposed, the concept of Pitman efficiency has to be extended to the sequential case. In the above mentioned paper by Lai (1978), a sequential analogon of the classical Pitman-Noether theorem was proved, which might be applied in the present context. Alternatively, we might take as our starting point a well-known result from the non-sequential theory which relates sample sizes with asymptotic shift parameters, cf. Witting-Nölle (1970), p. 161. The reasoning is completely parallel to the one given in Müller-Funk (1979).

In either case we arrive at the following definition. Let ψ, $\check{\psi} \in \mathcal{W}_{h_1,h_2}$ with asymptotic parameters ζ, $\check{\zeta}$, then we put

$$\text{ARE}\,(\psi : \check{\psi} \,|(h_1,h_2)) := (\zeta/\check{\zeta})^2.$$

In particular, if ψ and $\check{\psi}$ are the SCRT with score functions g_1,g_2 respectively $\check{g}_1,\check{g}_2$ then we obtain

$$\text{ARE}(\psi : \check{\psi} \,|(h_1,h_2)) = \left[\frac{\int_0^1 g_1(t)h_1(t)dt \cdot \int_0^1 g_2(t)h_2(t)dt}{\int_0^1 \check{g}_1(t)h_1(t)dt \cdot \int_0^1 \check{g}_2(t)h_2(t)dt}\right]^2 . \qquad (3.14)$$

Of course, these expressions are familiar from the non-sequential theory (cf. Behnen (1972)), i.e. the situation encountered there remains unaltered in the sequential set-up.

The possibility that the sequential tests treated so far may require an intolerably large number of observations will render them unacceptable in a variety of practical situations. In such cases we have to allow for some truncation or, alternatively, to choose a different type of stopping rule. Our results from section 2, however, will apply to restricted sequential tests as well. We refer to Sen (1978) as for such a type of restricted procedures.

Again in other situations, a quantitative statement on the degree of dependence may be preferred to a qualitative one. Suppose that for the correlation coefficient we want to construct a confidence interval with prescribed coverage probability β, $\beta = 1 - 2\alpha$, $0 < \alpha < 1/2$. Obviously, such a random interval has to be small in order to be meaningful, whence it is apparent that such a procedure will inevitably be sequential. It is near at hand to proceed along the lines suggested by Chow and Robbins (1965). On the grounds of invariance considerations or because of other reasons, we moreover demand this procedure to be nonparametric. If we conjecture that F_o (resp. G_o) represents the type of distribution which $L(X_1)$ (resp. $L(Y_1)$) belongs to, we shall choose g_1 (resp. g_2) to be the standardized inverse of F_o (resp. G_o). Note that,

$$\mathrm{corr}_H(X_1,Y_1) = \iint g_1(F_o)g_2(G_o)dH,$$

for every $H \in \mathcal{H}$ with marginals F_o,G_o. (Of course, it has been tacitely assumed that under F_o,G_o moments up to order two exist.)

This way we are lead to construct confidence intervals for $\mu(H)$ which are built upon $T_n = T_{n\,g_1g_2}$ where now g_1,g_2 are any non-decreasing score functions satisfying conditions (C). Let u_α be the upper α fractile of the standard normal law, choose $\theta > 0$ and define

$$\hat{\sigma}_n^2 := \sigma^2(H_n^*) + \frac{1}{n}, \quad n \in \mathbb{N}$$

$$I_n(\theta) := [\max\{T_n - \theta, -1\}, \min\{T_n + \theta, 1\}], \quad n \in \mathbb{N}$$

$$\tilde{N}(\theta) := \min(n \geqq 1 : \hat{\sigma}_n^2 \leqq \frac{\theta^2 n}{u_\alpha^2}\}.$$

In case, that under every fixed $H \in \mathcal{H}$ with $\sigma^2(H) > 0$ it holds true that

a) $(\hat{\sigma}_n^2)_{n \geqq 1}$ is a strongly consistent estimator for $\sigma^2(H)$,

b) $\{\theta^2 \tilde{N}(\theta) : 0 < \theta < \theta_o\}$ is uniformly integrable for some $\theta_o > 0$,

then the following statements (A), (B) can be verified by means of theorem 2.4.:

(A) $\lim_{\theta \to 0} P_H(\mu(H) \in I_{\tilde{N}(\theta)}(\theta)) = \beta$ ("asymptotic consistency")

(B) $\lim_{\theta \to 0} \dfrac{\theta^2 E_H(\tilde{N}(\theta))}{u_\alpha^2 \sigma^2(H)} = 1$ ("asymptotic efficiency").

Details will be given elsewhere.

REFERENCES

Behnen, K. (1972). A characterization of certain rank-order tests with bounds for the asymptotic relative efficiency. Ann. Math. Statist. 43, 1839-1851.

Bhuchongkul, S. (1964). A class of nonparametric tests for independence in bivariate populations. Ann. Math. Statist. 35, 138-149.

Bönner, N. (1976). Sequentielle Korrelationsrangtests, Unpublished Ph.D. thesis, Freiburg.

Chernoff, H. and Savage, I.R. (1958). Asymptotic normality and efficiency of certain nonparametric test statistics. Ann. Math. Statist. 29, 972-944.

Chow, Y. and Robbins, H. (1965). On the asymptotic theory of fixed width confidence intervals for the mean. Ann. Math. Statist. 36, 457-462.

Dvoretzky, A., Kiefer, J., Wolfowitz, J. (1953). Sequential decision problems for processes with continuous time parameter: testing hypotheses. Ann. Math. Statist. 24, 254-264.

Ghosh, M. (1972). On the representation of linear functions of order statistics. Sankhya, Series A, 34, 349-356.

Hoeffding, W. (1963). Probability inequalities for sums of bounded random variables. J. Amer. Statist. Assoc. 58, 13-30.

Lai, T.L. (1975a). On Chernoff-Savage statistics and sequential rank tests. Ann. Statist. 3, 825-845.

Lai, T.L. (1975b). A note on first exit time with applications to sequential analysis. Ann. Statist. 3, 999-1005.

Lai, T.L. (1978). Pitman efficiencies of sequential tests and uniform limit theorems in nonparametric statistics. Ann. Statist. 6, 1027-1047.

Lehmann, E. (1966). Some concepts of dependence. Ann. Math. Statist. 37, 1137-1153.

Müller-Funk, U. (1979). Nonparametric sequential tests for symmetry. Z. Wahrscheinlichkeitstheorie und Verw. Gebiete 46, 325-342.

Rüschendorf, L. (1977). Consistency of estimators for multivariate density functions and for the mode. Sankhya, Series A 39, 243-250.

Ruymgaart, F.H. (1974). Asymptotic normality of nonparametric tests for independence. Ann. Statist. 2, 892-910.

Ruymgaart, F.H., Shorack, G.R. and van Zwet, W.R. (1972). Asymptotic normality of nonparametric tests for independence. Ann. Math. Statist. 43, 1122-1135.

Sen, P.K. (1978). Nonparametric repeated significance tests. Developments in Statistics I, Krishnaiah, P.R. ed., Academic Press, New York, 227-265.

Sen, P.K. and Ghosh, M. (1971). On bounded length sequential confidence intervals based on one-sample rank order statistics. Ann. Math. Statist. 42, 189-203.

Sen, P.K. and Ghosh, M. (1974). Some invariance principles for rank statistics for testing independence. Z. Wahrscheinlichkeitstheorie und Verw. Gebiete 29, 93-107.

Tran, L.T. (1978). Some invariance principles for linear rank statistics for testing independence in bivariate populations. Unpubl. manuscript, Indiana University.

Wegman, E. (1971). A note on the estimation of the mode. Ann. Math. Statist. 42, 1909-1915.

Whitt, W. (1970). Weak covergence of probability measures on the function space $C[0,\infty)$. Ann. Math. Statist. 41, 939-944.

Witting, H. and Nölle, G. (1970). Angewandte Mathematische Statistik, Teubner, Stuttgart.

WIENER-LÉVY MODELS, SPHERICALLY EXCHANGEABLE TIME SERIES AND SIMULTANEOUS INFERENCE IN GROWTH CURVE ANALYSIS

C. B. Bell*
F. Ramirez
Eric Smith

Department of Biostatistics
University of Washington
Seattle, Washington

INTRODUCTION AND SUMMARY

The object of this treatise is the development of statistical methodology and first-order modelling for some common types of biomedical time series. These series arise in two distinct manners. Continuous parameter data, e.g., albumen levels in postoperative cancer patients, can feasibly only be recorded at discrete times, and the resultant data form time series. Other data such as total caloric intake is taken on a daily basis and naturally constitute time series.

The authors recognize that any long-term serious study of such phenomena should be based on a thorough study of the physiological background of the underlying processes (in conjunction with experts in the field), as well as a statistical analysis of the data. That sort of study is, perhaps, best exemplified by the works of Tautu and Iosefscu (1973 (a), (b)).

This short treatise will not attempt to bring in the physiological background. Instead, it will begin with a simple one-ancillary parameter

*Research partially supported by Biomedical Research Support Grant (BRSG) #5S01 RR 05714, at the University of Washington.

ISBN: 0-12-166650-6.

model, and enlarge the parameter space by considering Bayesian priors on that parameter. The idea here is that a large enough parameter space should accommodate a variety of types of data sets.

A preliminary investigation of the available data sets indicates that (a) within-patient data is dependent; and (b) there is independence between patients. An initial model which has these two features associates with the ith patient: $Z_i(t) = \mu(t) + \sigma Z_i^*(t)$, where σ is the unknown ancillary parameter; $\mu(\cdot)$ is the deterministic mean function (or growth curve) of the "ensemble" of patients; and $\{Z^*(t)\}$ is an appropriate Wiener-Lévy stochastic process. This model introduces one level of dependence beyond the classical k-sample statistical model; and as such is amenable to many useful techniques.

The methodology to be developed concerns confidence intervals and tests about the mean function $\mu(\cdot)$; and will be based on the simultaneous inference techniques related to the Studentized Range (S-R) statistic; the Maximum Modulus (M-M) statistic; and the linear contrasts theory of Scheffé. (See, e.g., Miller, 1977.) The actual data sets and their analyses will be presented elsewhere. Only specific confidence intervals and test statistics and some associated robustness questions will be presented in the sequel.

In Section I, a detailed description of the initial model and the inference problems are given. A basic data transformation and the fundamental simultaneous inference tools are given in Section II. The pertinent confidence intervals and tests are presented in Sections III and IV, respectively. Section V introduces the S-E (spherical exchangeability) property, which is incorporated into the Bayesian model of Section VI. Conclusions and open problems constitute Section VII.

I. THE BASIC MODEL AND ASSOCIATED INFERENCE PROBLEMS

It is assumed that each of "r" patients is represented by a stochastic process $\{Z_i(t): t \geq 0\}$, where, for all i,

(a) $Z_i(t) = \mu(t) + \sigma Z_i^*(t)$;

(b) $Z_i^*(0) = 0$;

(c) $E(Z_i^*(t)) = 0$; and

(d) $\mathrm{Cov}(Z_i^*(s), Z_i^*(t)) = \min(s,t)$.

Further (e) the Wiener-Lévy processes $[\{Z_i(t): t > 0\}: 1 \leq i \leq r]$ are statistically independent.

For a variety of practical and theoretical reasons, it will be assumed that the continuous parameter processes above are all sampled at times: $\Delta, 2\Delta, \ldots, c\Delta$. [For the real data on which this analysis is based, Δ most usually is 24 hours.]

The problem here is to gain as much information as possible about the mean function $\mu(\cdot)$, from the (r x c) data matrix $\tilde{Z} = (z_{ij})$, where $z_{ij} = Z_i(j\Delta)$.

On the basis of this data matrix, one can then make inference about

(I) the mean vector $[\mu(\Delta),\ldots,\mu(c\Delta)]$ and its components;

(II) μ^*, when $\mu(t) \equiv \mu^*$; and

(III) α and β, when $\mu(t) = \alpha + \beta t$.

[Of course, other functional forms are somewhat amenable to the methodology to be developed below, but they will not be treated in this paper .]

II. DATA TRANSFORMATION AND SIMULTANEOUS INFERENCE

The covariance structure of the model entails independent increments, and one is lead naturally to the transformed data matrix below.

$\tilde{X} = \tilde{Z}\tilde{A}$ where $\tilde{A} = (a_{ij})$ and $a_{ij} = \begin{cases} 1 \text{ if } i=j \\ -1 \text{ if } j=i+1 \\ 0 \text{ otherwise,} \end{cases}$. Here, $\tilde{X} = (X_{ij})$

with $X_{i1} = Z_i(\Delta)$; and $X_{ij} = Z_i(j\Delta) - Z_i([j-1]\Delta)$, for $j \geq 2$. If one defines $\mu_j = \begin{cases} \mu(\Delta) \text{ for } j=1 \\ \mu(j\Delta) - \mu([j-1]\Delta) \text{ for } j \geq 2, \end{cases}$ then the parameters for the elements of the transformed data matrix $\tilde{X}$ are as given in Table 2.1.

TABLE 2.1

	μ_1	μ_2		μ_c
$\sigma^2\Delta$	X_{11}	X_{12}		X_{1c}
. . .				
$\sigma^2\Delta$	X_{r1}	X_{r2}		X_{rc}

More precisely, one has

Lemma 2.1 - $\{\frac{X_{ij}-\mu_j}{\sigma\sqrt{\Delta}}: 1 \leq i \leq r, 1 \leq j \leq c\}$ are i.i.d. $N(0,1)$.

Standard simultaneous inference concerning $(\mu_1,\ldots,\mu_c)$ is embodied in the theorem below.

Let $F(m; k; \cdot)$, $S\text{-}R(m; k; \cdot)$ and $M\text{-}M(m; k; \cdot)$ be the cpfs of the F; Studentized Range; and Maximum Modulus distributions, respectively, with the indicated degrees of freedom; and let $S_c^2 = q^{-1} \sum_i \sum_j (X_{ij}-\bar{X}_{.j})^2$ where $q = c(r-1)$.

<u>Theorem 2.2</u>

(i) $P\{\sum_1^c a_j\mu_j \;\varepsilon\; \sum_1^c a_j\overline{X}_{.j} \pm S_c\sqrt{\gamma}\,\sqrt{\sum a_j^2}\,\sqrt{\frac{c}{r}}$ for all $(a_1,\ldots,a_c)\}$

$= F(c;\, q;\, \gamma)$;

(ii) $P\{\sum_1^c b_j\mu_j \;\varepsilon\; \sum_1^c b_j\overline{X}_{.j} \pm S_c\gamma r^{-\frac{1}{2}}\, 2^{-1}\sum_1^c |b_j|$ for all $(b_1,\ldots,b_c)$

with $\sum b_j = 0\} = S\text{-}R(c;\, q;\, \gamma)$.

(iii) $P\{\sum_1^c d_j\mu_j \;\varepsilon\; \sum_1^c d_j\overline{X}_{.j} \pm S_c\gamma r^{-\frac{1}{2}}\sum_1^c |d_j|$ for all $(d_1,\ldots,d_c)\}$

$= M\text{-}M(c;\, q;\, \gamma)$.

However, the prime interest here is in $[\mu(\Delta),\ldots,\mu(c\Delta)]$ rather than $[\mu_1,\ldots,\mu_c]$; and one must "convert" the theorem above using the facts below.

(A) $\mu(j\Delta) = \mu_1 + \ldots + \mu_j$

(B) $\sum_1^c \eta_j\mu(j\Delta) = \sum_{j=1}^c \left(\sum_{m=j}^c \eta_m\right)\mu_j$

(C) $\overline{Z}(j\Delta) = \overline{X}_{.1} + \ldots + \overline{X}_{.j}$

(D) $\sum_1^c \overline{Z}(j\Delta)\eta_j = \sum_1^c \left(\sum_{m=j}^c \eta_m\right)\overline{X}_{.j}$

The "converted" theorem is now

<u>Theorem 2.3</u>

(i) $P\{\sum_1^c \alpha_j\mu(j\Delta) \;\varepsilon\; \sum_1^c \alpha_j\overline{Z}(j\Delta) \pm S_c\sqrt{\gamma}\,\sqrt{\sum_{j=1}^c \left(\sum_{m=j}^c \alpha_m\right)^2}\,\sqrt{\frac{c}{r}}$

for all $(\alpha_1,\ldots,\alpha_c)\} = F(c;\, q;\, \gamma)$.

(ii) $P\{\sum_{1}^{c} \beta_j \mu(j\Delta) \varepsilon \sum_{1}^{c} \beta_j \overline{Z}(j\Delta) \pm S_c \gamma 2^{-1} r^{-\frac{1}{2}} \sum_{j=1}^{c} | \sum_{m=j}^{c} \beta_m |$

for all $(\beta_1,\ldots,\beta_c)$ with $\sum_{j=1}^{c} j\beta_j = 0\} = $ S-R(c; q; γ).

(iii) $P\{ \sum_{j=1}^{c} \eta_j \mu(j\Delta) \varepsilon \sum_{j=1}^{c} \eta_j \overline{Z}(j\Delta) \pm S_c (\sum_{j=1}^{c} | \sum_{m=j}^{c} \eta_m |) \gamma r^{-\frac{1}{2}}$

for all $(\eta_1,\ldots,\eta_c)\}= $ M-M(c; q; γ).

From the theorem above, it is seen that simultaneous inference based on the S-R statistic is limited to certain restricted contrasts, and, hence, will not apply to many of the contrasts of interest. The Scheffé and M-M results will apply to all of these contrasts.

In most of the cases below, the number of contrasts is finite. Hence, the inference is conservative. Further, when the concern is for a single contrast one uses a t-statistic rather than Theorem 2.2.

III. CONFIDENCE INTERVALS

The intervals of "natural" interest in this section do not satisfy the "converted" conditions on S-R contrasts. Hence, that statistic will not be used in this section. It, of course, could be used if one were interested in contrasts satisfying the restrictions.

Let f', v' satisfy $F(c; q; f') = 1 - \alpha$; and M-M$(c; q; v') = 1 - \alpha$, respectively, and let t' be the $100(1 - \frac{\alpha}{2})$th percentile of the t-distribution with q degrees of freedom. (Recall $q = c(r-1$.)

Then, for the components of the mean vector, one has

Theorem 3.1

Both of the events below have probabilities $\geq 1 - \alpha$.

(i) $\{[\mu(\Delta),\ldots,\mu(c\Delta)] \in \bigotimes_{1}^{c}[\overline{Z}(j\Delta) \pm S_c \sqrt{\frac{cjf'}{r}}]\}$.

(ii) $\{[\mu(\Delta),\ldots,\mu(c\Delta)] \in \bigotimes_{1}^{c}[\overline{Z}(j\Delta) \pm \frac{S_c jv'}{\sqrt{r}}]\}$.

For differences of components of the mean vector it follows that

Theorem 3.2

Both of the events below have probabilities $\geq 1 - \alpha$.

(i) $\{\mu(s\Delta) - \mu(m\Delta) \in \overline{Z}(s\Delta) - \overline{Z}(m\Delta) \pm S_c \sqrt{f'c|s-m|r^{-1}}$ for all s and $m\}$.

(ii) $\{\mu(s\Delta) - \mu(m\Delta) \in \overline{Z}(s\Delta) - \overline{Z}(m\Delta) \pm \frac{S_c v'|s-m|}{\sqrt{r}}$ for all s and $m\}$.

As previously mentioned, single contrasts involve only the t-distribution. The principal result here is as follows.

Theorem 3.3

The events below each have probability $= 1 - \alpha$.

(i) $\{\mu(s\Delta) \in \overline{Z}(s\Delta) \pm t'S_c\sqrt{sr^{-1}}\}$.

(ii) $\{\mu(s\Delta) - \mu(m\Delta) \in \overline{Z}(s\Delta) - \overline{Z}(m\Delta) \pm t'S_c\sqrt{|s-m|r^{-1}}\}$.

(iii) $\{\sum_{1}^{c} d_j\mu(j\Delta) \in \sum_{1}^{c} d_j\overline{Z}(j\Delta) \pm t'S_c\sqrt{\sum_{j=1}^{c}(\sum_{m=j}^{c} d_m)^2 r^{-1}}\}$.

Finally, if one assumes that $\mu(\cdot)$ has a simple functional form, it is possible to give confidence intervals for certain parameters.

Theorem 3.4

Each of the events below have probability = 1 - α under the indicated circumstances.

(i) $\{\mu^* \varepsilon \bar{Z}(\Delta) \pm t'S_c r^{-\frac{1}{2}}\}$ (when $\mu(t) \equiv \mu^*$).

(ii) $\{\alpha \varepsilon \frac{c\bar{Z}(\Delta)-\bar{Z}(c\Delta)}{c-1} \pm t'S_c\sqrt{\frac{c}{(c-1)r}}\}$ (when $\mu(t) = \alpha + \beta t$).

(iii) $\{\beta\varepsilon[\bar{Z}(c\Delta) - \bar{Z}(\Delta)][\Delta(c-1)]^{-1} \pm t'S_c\Delta[r(c-1)]^{-\frac{1}{2}}\}$

(when $\mu(t) = \alpha + \beta t$).

Corresponding to each of the confidence intervals above there is, of course, an appropriate testing situation. These are given below, as are some new situations.

IV. TESTS OF HYPOTHESES

Table 4.1 gives the test statistics and their null distributions for the indicated hypotheses corresponding to confidence intervals of the preceding section.

TABLE 4.1

Hypothesis	Test Statistic	Null Distribution
1. $\mu(t) \equiv \mu^*$	$[\bar{Z}(\Delta) - \mu^*]\sqrt{rS_c}^{\,-1}$	t_q
2. $\mu(t) = \alpha^* + \beta^* t$	$\dfrac{r\{\sum_2^c[\bar{X}_{\cdot j} - \beta^*\Delta]^2 + [\bar{Z}(\Delta) - \alpha^* - \beta^*\Delta]^2\}}{cS_c^2}$	$F(c,q)$
3. $\mu(s\Delta) = a^*$	$[\bar{Z}(s\Delta) - a^*]S_c^{-1}\sqrt{rs^{-1}}$	t_q
4. $\mu(s\Delta) - \mu(m\Delta) = b^*$	$\dfrac{[\bar{Z}(s\Delta) - \bar{Z}(m\Delta) - b^*]\sqrt{r}}{S_c\sqrt{\lvert s-m\rvert}}$	t_q
5. $\sum_1^c d_j\mu(j\Delta) = \gamma^*$	$\dfrac{[\sum_1^c d_j\bar{Z}(j\Delta) - \gamma^*]\sqrt{r}}{S_c\sqrt{\sum_1^c(\sum_{j=m}^c d_m)^2}}$	t_q
6. $[\mu(\Delta), \ldots, \mu(c\Delta)] = [a_1^*, \ldots, a_c^*]$	$\dfrac{r\sum_1^c[\bar{X}_{\cdot j} - a_j^* + a_{j-1}^*]^2}{cS_c^2}$ (where $a_o = 0$)	$F(c,q)$

Two additional hypothesis of interest are not related to the confidence intervals in the same fashion as those above. These are given in Table 4.2.

TABLE 4.2

Hypothesis	Test Statistic	Null Distribution
$\mu(t)$ is constant	$\sum_{j=2}^{c} (\overline{X}_{.j})^2 \, S_c^{-2} c^{-1} \, r$	$F(c-1,q)$
$\mu(t)$ is linear	$r\sum_{2}^{c}(\overline{X}_{.j}-X^*)^2 S_c^{-2}(c-2)^{-1}$ where $X^* = (c-1)^{-1} \sum_{2}^{c} \overline{X}_{.j}$	$F(c-2,q)$

It is possible to treat the optimality of many of the tests and confidence intervals above in the standard fashion. This will not be done in this article. What will be treated in this article is an extension of the above results to the case of spherically distributed errors.

V. S-E TIME SERIES

It seems essential to the theorems of the preceding two sections that the X_{ij} be independent and normal. These conditions are sufficient but not necessary, as will be seen in the sequel.

Definition 5.1 - A time series $\{Y_n: n \geq 1\}$ is S-E if for each k, each $i_1 < \ldots < i_k$ and orthogonal matrix $\tilde{C}_k$, $(Y_1,\ldots,Y_k)' \stackrel{d}{=} \tilde{C}_k(Y_{i_1},\ldots,Y_{i_k})'$.

It is clear that each of the following three examples satisfies the definition.

<u>Example 5.1</u> $Y_1, \ldots, Y_n, \ldots,$ i.i.d. $N(0,\sigma^2)$.

<u>Example 5.2</u> $(Y_1, \ldots, Y_{15})'$ distributed uniformly over $\{2 \le \sum_1^{15} y_j^2 \le 5\}$.

<u>Example 5.3</u> $Y_1, \ldots, Y_n, \ldots,$ are conditionally i.i.d. $N(0,W)$ and $W \sim J(\cdot)$ with $J(0) = 0$.

From the last example, one can obtain an F-distribution as follows.

<u>Example 5.4</u> Consider Example 5.3 with $J(w) = 1-e^{-w}$, for $w>0$. Then

$$P\left\{\frac{\sum_1^4 Y_j^2}{\sum_5^{12} Y_j^2}(2) \le z \,\middle|\, W = w\right\} = F(4;\ 8;\ z) \text{ if } w \ne 0.$$ Hence, unconditionally

$$\left[\sum_1^4 Y_j^2\right]\left[\sum_5^{12} Y_j^2\right]^{-1}(2) \sim F(4;\ 8).$$

Lord (1954) and Kelker (1970) have indicated that Example 5.4 can be generalized as follows.

<u>Theorem 5.1</u>

Let $\{Y_n\}$ be a finite or infinite S-E time series. Then

$$\left[\sum_1^k Y_j^2\right]\left[\sum_{k+1}^m Y_j^2\right]^{-1} k(m-k)^{-1} \sim F(k;\ m-k).$$

Further, Schoenberg (1938)has essentially shown that Example 5.3 can be generalized.

Theorem 5.2

Let $\{X_n\}$ be an infinite S-E time series. Then there exists a cpf $J(\cdot)$ with $J(0) = 0$ such that for all n and for all $x_1,\ldots,x_n$

$$P\{X_1 \le x_1; \ldots ; X_n \le x_n\} = \int_0^\infty \prod_1^n \Phi(x_j w^{-\frac{1}{2}}) dJ(w).$$

This means that an infinite S-E time series is always a variance mixture of i.i.d. 0-mean normals. This result is not valid in the finite case as is seen by Example 5.2.

However, we can show

Theorem 5.3

For both finite and infinite S-E time series, $\underset{\sim}{Y} = (Y_1,\ldots,Y_k)'$

(i) has mean $\underset{\sim}{0}$ if first moments exist; and

(ii) has a covariance matrix which is a constant multiple of the identity matrix of order k, i.e., $a\tilde{I}_k$, if second moments exist.

Of course, due to Theorem 5.2, one has in the infinite case a more detailed conditional structure.

Theorem 5.4

Let $\{Y_n\}$ be an infinite S-E time series. Then, for each k and $\underset{\sim}{Y} = (Y_1,\ldots,Y_k)'$ (given W of Theorem 5.2)

(i) the conditional mean vector exists and $= \underset{\sim}{0}$,

(ii) the conditional covariance matrix exists and is of the form $a\tilde{I}_k$, as above; and

(iii) $\underset{\sim}{Y}$ is conditionally normal

These latter two results will be of interest in the next section. However, Theorems 5.1 and 5.2 are useful for establishing the following result.

Theorem 5.5

Let $\{X_{ij} - \mu_j: 1 \le i \le r, 1 \le j \le c\}$ be S-E. Then all of the results (i.e., confidence intervals and tests) of Sections III and IV are valid.

That is to say, the "errors" need not be i.i.d. It is sufficient to have the totality of errors for the set of patients to be S-E.

One now seeks a simple formulation in terms of individual patients which will achieve the S-E property exactly or approximately.

VI. A BAYESIAN MODEL AND A QUESTION OF ROBUSTNESS

A preliminary analysis of some of the real data indicates that in several cases it is reasonable to assume, besides (a) and (b) of the introduction.

(c) that the within-patient variability varies from patient to patient.

These assumptions lead to the following initial modification of the original model.

(*) Let $\{W_1,\ldots,W_r\}$ and $\{[Z_j^*(t): t \ge 0], j = 1,2,\ldots,r\}$ be statistically independent, with $\{Z_j^*(t)\}$ distributed as in Section II; and $W_1,\ldots,W_r$ i.i.d. $J(\cdot)$, and $J(0) = 0$. Let $Z_j(t) = \mu(t) + \sqrt{W_j}\, Z_j^*(t)$. [One notes here that $J(\cdot)$ can be considered as a prior distribution in the Bayesian sense.]

On taking differences as in Table 2.1, i.e., $\tilde{X} = \tilde{Z}\tilde{A}$, then one obtains Table 6.1, where the μ_j's are as previously defined and the $W_i\Delta$ are the conditional variances.

TABLE 6.1

	μ_1	μ_2		μ_c
$W_1\Delta$	X_{11}	X_{12}		X_{1c}
$W_2\Delta$	X_{21}	X_{22}		X_{2c}
$\vdots$				
$W_r\Delta$	X_{r1}	X_{r2}		X_{rc}

Corresponding to Lemma 2.1, one proves

Lemma 6.1 - (i) $\{\frac{X_{ij}-\mu_j}{\sqrt{W_i\Delta}}: 1 \le i \le r, 1 \le j \le c\}$ are conditionally i.i.d. given $\underset{\sim}{W} = (W_1,\ldots,W_r)$.

(ii) $\{[X_{r_1}-\mu_1,\ldots,X_{i_c}-\mu_c], i = 1,2,\ldots,r\}$ are i.i.d. S-E time series.

(iii) $\{X_{ij} - \mu_i: 1 \le i \le r, 1 \le j \le c\}$ is NOT S-E.

The third conclusion of the above lemma tells one that the results of Sections III and IV do not apply exactly.

In order to achieve the appropriate robustness, one might try the technique below.

For illustrative purposes, one will start with the case of four patients and two times, i.e., r = 4, and c = 2.

Let $\tilde{D}_4 = \frac{1}{4}\begin{pmatrix} 1 & 1 & 1 & 1 \\ 1 & 1 & -1 & -1 \\ 1 & -1 & 1 & -1 \\ 1 & -1 & -1 & 1 \end{pmatrix} = (d_{ij})$ and

$$\tilde{L}=\tilde{D}_4\tilde{X}, \text{ where } \tilde{X} = \begin{pmatrix} X_{11} & X_{12} \\ \vdots & \vdots \\ X_{41} & X_{42} \end{pmatrix} = \begin{pmatrix} Z_1(\Delta) & Z_1(2\Delta) - Z_1(\Delta) \\ \vdots & \vdots \\ Z_4(\Delta) & Z_4(2\Delta) - Z_4(\Delta) \end{pmatrix}.$$

Definition 6.1 - (a) Functions $L_i(.,.,.,.)$ are defined by

$$\tilde{L} = \begin{pmatrix} L_{11} & L_{12} \\ \vdots & \vdots \\ L_{41} & L_{42} \end{pmatrix} = \begin{pmatrix} L_1(X_{11},\ldots,X_{41}) & L_1(X_{12},\ldots,X_{42}) \\ & \\ L_4(X_{11},\ldots,X_{41}) & L_4(X_{12},\ldots,X_{42}) \end{pmatrix}$$

Further, (b) $\underset{\sim}{T} = (L_{11},\ldots,L_{41},L_{12},\ldots,L_{42})'$.

Question - How close does $\underset{\sim}{T}$ come to satisfying the S-E property?

It turns out that (i) and (ii) of Theorem 5.3 are satisfied, as are (i) and (iii) of Theorem 5.4. However, Theorem 5.4 (ii) is only satisfied in a restricted asymptotic sense.

More precisely, one has

Theorem 6.2

Assume second moments exist. Then

(i) $E(\underset{\sim}{T}) = \underset{\sim}{0}$; and

(ii) $\tilde{\Sigma}_{\underset{\sim}{T}} = a\tilde{I}_8$, where a is a positive constant and $\tilde{I}_8$ is the unit matrix in 8-space.

Further, conditionally, given $\underset{\sim}{W}^{(4)} = (W_1,\ldots,W_4)$

(iii) $\underset{\sim}{T}$ has a multinormal distribution;

(iv) with mean $\underset{\sim}{0}$; and

(v) with covariance matrix of the

$$\tilde{\Sigma}_{\underset{\sim}{T}}(\underset{\sim}{W}^{(4)}) = \frac{1}{4}\begin{pmatrix} \tilde{B} & \tilde{0} \\ \tilde{0} & \tilde{B} \end{pmatrix} = \left[\sigma_{ij}(\underset{\sim}{W}^{(4)})\right]$$

Here $\tilde{B}$ is a 4 x 4 matrix with

(a) diagonal elements equal to $\tilde{W} = \frac{1}{4}(W_1+\ldots+W_4)$ and

(b) each off diagonal element is either $L_2(\underset{\sim}{W})$, $L_3(\underset{\sim}{W})$ or $L_4(\underset{\sim}{W})$

where the L's are defined above.

Finally (vi) $\underset{\sim}{T}$ is NOT S-E.

The impediment to attaining the S-E property is the fact that the off-diagonal elements of $\tilde{B}$ do not vanish.

These off-diagonal elements do have the interesting property given below.

Lemma 6.3 - $4\sigma_{ij}(\underset{\sim}{W}^{(4)}) \stackrel{d}{=} \frac{W_1+W_2}{4} - \frac{W_3+W_4}{4}$ for each $i \neq j$.

This idea, then, gives one an asymptotic result of some importance.

Consider $\{\tilde{D}_r^*: \ r = 4, 8,16,32,\ldots\}$ as used in 2^m-factorial designs, and let $\tilde{D}_r = \frac{1}{r}\tilde{D}_r^*$, as was done above for $r = 4$. With this framework, one has the following result which is related to the robustness question.

Theorem 6.4

Assume second moments exist; and consider $\underset{\sim}{T}^{(r)}$ appropriately related to $\tilde{D}_r\tilde{X}$ for c fixed and $r = 4,8,16,\ldots$, and $\underset{\sim}{W}^{(r)} = (W_1,\ldots,W_r)$. Then for the conditional covariance matrix of $\underset{\sim}{T}^{(r)}$, one has

(i) $r\sigma_{ij}$ tends in probability to $\mu_J = \int_0^\infty w\,dJ(w)$; and

(ii) $r\sigma_{ij}$ (for $i \neq j$) tends to 0 in probability.

On the basis of this last result and experience with the statistics used in the inference, it is reasonable to <u>suspect</u> that the procedures above are robust as applied to $\{L_{ij}:\ 1 \le i \le r,\ 1 \le j \le c\}$. Here,

$$L_{1j} = \overline{X}_{\cdot j} \quad \text{and} \quad S_c^2 = Q = 4q^{-1} \sum_{i=2}^{r} \sum_{j=1}^{c} L_{ij}^2$$ and the results

of Sections III and IV should hold approximately.

VII. CONCLUSIONS AND OPEN PROBLEMS

From the preceding development, one makes several conclusions.

(A) The methodology of Sections III and IV holds exactly for the initial model $Z_j(t) = \mu(t) + \sigma Z_j^*(t)$.

(B) The techniques are <u>probably</u> (in a Bayesian sense) robust for the case of variances "approximately" equal.

(C) Conclusion (B) holds for the Bayesian model: $Z_j(t) = \mu(t) + \sqrt{W_j}\, Z_j^*(t)$ when $J(b+\varepsilon) - J(b) = 1$ for sufficiently "small" positive ε.

There are, of course, more open problems here than there are solutions. These involve specific questions of robustness and specific modifications of the model.

(1) <u>Robustness</u> - What specific robustness results can be established? What happens when $J(\cdot)$ is gamma, or chi, or Pareto? What is the difference between the "performances" for thick-tailed and thin-tailed distributions?

(2) <u>Extension to Other Classical Frameworks</u> - All of the procedures presented here refer to one-sample inference. In actual practice, one is often concerned with groups of patients exposed to different treatments. How can this methodology be extended to the 2-sample and k-sample cases?

(3) Sampling Rates - The procedures above were developed for a uniform sampling rate, i.e., one observation per patient per unit time. In actual practice this may not be possible or desirable. How can the procedures be modified to handle non-constant sampling rates? What are some optimal sampling plans?

(4) Missing Data - In practice, of course, a non-trivial portion of the data is often missing. This situation has not been touched on above. What modifications are necessary to handle missing data?

(5) Parametric Regression - The cases of a general mean function as well as constant and linear mean functions were treated above. However, on the basis of the data and other considerations, it would be desirable to consider polynomial and other non-linear parametric functional forms for $\mu(\cdot)$. Which forms can be adequately handled by the procedures developed above?

(6) A Mixed Model - Some of the discussion of the data suggests that

(a) some patients remain above (or below) the ensemble average; and

(b) their variabilities are approximately proportional to this deviation. This suggests a model of the form $Z_j(t) = \mu(t) + V_j + \sigma V_j Z_j^*(t)$ where the V's are random variables. How does one treat this mixed Bayesian-homoscedastistic model?

ACKNOWLEDGMENTS

The authors wish to thank Nancy Flournoy for making available data on which the techniques of Section III were applied. The authors also wish to thank Joel Brodsky whose critical reading of an initial draft led to a complete rewriting of several sections.

REFERENCES

Bell, C.B.; Woodroofe, M.F.; Avadhani, T.V. (1970). "Nonparametric inference for stochastic processes" in Nonparametric Techniques in Statistical Inference, Ed. by Madan Lal Puri, Cambridge University Press.

Bell, C.B. (1975). "Circularidad en Estadistica," Trabajos de E. y de Inv. Op., 26, 61-81.

Iosifescu, M. and Tautu, P. (1973). Stochastic Processes and Applications in Biology and Medicine I. Theory. Springer-Verlag, New York.

Iosifescu, M. and Tautu, P. (1973). Stochastic Processes and Applications in Biology and Medicine II. Models. Springer-Verlag, New York.

Kelker, Douglas (1970). "Distribution theory of spherical distributions and a location-scale parameter generalization," Sankhya: Series A, 32, 419-430.

Kelker, Douglas (1971). "Infinite divisibility and variance mixtures of the normal distribution," Ann. Math. Statist., 42, 824-827.

Lord, R.D. (1954). "The use of Hankel transforms in statistics. I. General theory and examples," Biometrika, 41, 44-55.

Miller, Rupert G., Jr. (1966). Simultaneous Statistical Inference, McGraw-Hill Book Co., New York.

Miller, Rupert G., Jr. (1977). "Developments in multiple comparisons. 1966-1976," JASA, 72, 779-788.

Schoenberg, I.J. (1938). "Metric spaces and completely monotone functions," Ann. Math., 39, 811-841.

A NOTE TO THE CHUNG-ERDÖS-SIRAO THEOREM

P. Révész

Mathematical Institute
Hungarian Academy of Sciences
Budapest, Hungary

Let $\{W(t);\ t \geq 0\}$ be a Wiener process and define the stochastic process

$$c_1(h) = \sup_{0 \leq t \leq 1-h} h^{-1/2}(W(t+h) - W(t)) \quad (0 \leq h \leq 1).$$

The Chung-Erdös-Sirao (1959) theorem investigates the question: How big can be the process $c_1(h)$ as $h \to 0$? In this paper the analogous question will be investigated: How small can be the process $c_1(h)$ as $h \to 0$?

I. INTRODUCTION

Let $\{W(t);\ t \geq 0\}$ be a Wiener process. The continuity properties of this process can be characterized by the following stochastic processes:

$$c_1(h) = \sup_{0 \leq t \leq 1-h} h^{-1/2}(W(t+h) - W(t)) \quad (0 \leq h \leq 1),$$

$$c_2(h) = \sup_{0 \leq t \leq 1-h} h^{-1/2}|W(t+h) - W(t)| \quad (0 \leq h \leq 1),$$

$$c_3(h) = \sup_{0 \leq t \leq 1-h} \sup_{0 \leq s \leq h} h^{-1/2}(W(t+s)-W(t)) \quad (0 \leq h \leq 1),$$

$$c_4(h) = \sup_{0 \leq t \leq 1-h} \sup_{0 \leq s \leq h} h^{-1/2}|W(t+s)-W(t)| \quad (0 \leq h \leq 1).$$

ISBN: 0-12-166650-6

The classical Lévy (1937) theorem states:

Theorem A. We have

$$\lim_{h\to 0} \frac{c_i(h)}{(2 \log h^{-1})^{1/2}} = 1 \quad \text{a.s.} \qquad (i = 1,2,3,4). \tag{1}$$

Chung-Erdös-Sirao (1959) (see also Pruitt-Orey, 1973) gave a deeper insight to see: how big can be the processes $c_i(h)$ $(i = 1,2,3,4)$. Their theorem implies

Theorem B. For any $\varepsilon > 0$ and for almost all $\omega\in\Omega$ (the basic space) there is a $0 < h_0 = h_0(\varepsilon,\omega) < 1$ such that

$$c_i(h) \le a_1(h) = (2 \log h^{-1}+(5+\varepsilon)\log \log h^{-1})^{1/2} \qquad (i = 1,2,3,4)$$

if $h \le h_0$.

Theorem C. For almost all $\omega\in\Omega$ there is a sequence $1 > h_1 = h_1(\omega) > h_2=h_2(\omega) > \ldots > 0$ such that

$$c_i(h_\ell) \ge a_2(h_\ell) = (2 \log h_\ell^{-1} + 5 \log \log h_\ell^{-1})^{1/2} \qquad (i = 1,2,3,4 \;;\; \ell = 1,2,\ldots).$$

Theorems B and C together clearly imply that

$$\limsup_{h\to 0} \frac{c_i(h)}{(2 \log h^{-1})^{1/2}} = 1 \quad \text{a.s.} \qquad (i = 1,2,3,4)$$

but they (or even their more precise forms) do not imply (1).

In this paper we intend to give the analogues of Theorems B and C answering the question: How small can be the process $c_1(h)$? Let us introduce the following definitions:

Definition 1. The function $\alpha_1(h)$ $(0 < h < 1)$ belongs to the upper-upper class of the process $c_i(h)$ $(\alpha_1 \in UUC(c_i))$ if for almost all $\omega \in \Omega$ there exists a $h_0 = h_0(\omega)$ such that $c_i(h) \le \alpha_1(h)$ for every $h \le h_0$.

Definition 2. The function $\alpha_2(h)$ $(0 < h < 1)$ belongs to the upper-lower class of the process $c_i(h)$ $(\alpha_2 \in ULC(c_i))$ if for almost all $\omega \in \Omega$ there exists a sequence $1 > h_1 = h_1(\omega) > h_2 = h_2(\omega) > \ldots > 0$ such that $c_i(h_\ell) \ge \alpha_2(h_\ell)$ $(\ell = 1,2,\ldots)$.

Definition 3. The function $\alpha_3(h)$ $(0 < h < 1)$ belongs to the lower-upper class of the process $c_i(h)$ $(\alpha_3 \in LUC(c_i))$ if for almost all $\omega \in \Omega$ there exists a sequence $1 > h_1 = h_1(\omega) > h_2 = h_2(\omega) > \ldots > 0$ such that $c_i(h_\ell) \le \alpha_3(h_\ell)$ $(\ell = 1,2,\ldots)$.

Definition 4. The function $\alpha_4(h)$ $(0 < h < 1)$ belongs to the lower-lower class of the process $c_i(h)$ $(\alpha_4 \in LLC(c_i))$ if for almost all $\omega \in \Omega$ there exists a $h_0 = h_0(\omega)$ such that $c_i(h) \ge \alpha_4(h)$ for every $h \le h_0$.

Theorems B and C say that $a_1(h) \in UUC(c_i))$ $(i = 1,2,3,4)$ and $a_2(h) \in ULC(c_i))$ $(i = 1,2,3,4)$. The Chung-Erdös-Sirao theorem gives a complete description of the monotone functions of the classes $UUC(c_i)$ and $ULC(c_i)$ $(i = 1,2,3,4)$.

Here we will investigate the classes $LUC(c_1)$ and $LLC(c_1)$ but we will have a non-complete description only. It is interesting to point out already here that the Chung-Erdös-Sirao theorem implies that if $\alpha_1(h)\in UUC(c_i)$ and $\alpha_2(h)\in ULC(c_i)$ and they are monotone functions then $\alpha_1^2(h) - \alpha_2^2(h) \to \infty$ as $h \to 0$. However for any $\varepsilon > 0$ we will give an $a_3(h,\varepsilon)\in LUC(c_i)$ and an $a_4(h,\varepsilon)\in LLC(c_1)$ such that $a_3^2(h,\varepsilon) - a_4^2(h,\varepsilon) \le \log(9+\varepsilon)$. A similar phenomenon was found by Erdös-Révész (1976) for coin tossing.

Now, we formulate our main results:

Theorem 1. For any $\varepsilon > 0$ and for almost all $\omega\in\Omega$ there exists a sequence $1 > h_1 = h_1(\omega,\varepsilon) > h_2 = h_2(\omega,\varepsilon) >\ldots> 0$ such that

$$c_1(h_\ell) \le a_3(h_\ell) \qquad (\ell = 1,2,\ldots) \tag{2}$$

where

$$a_3(h) = a_3(h,\varepsilon) = (2 \log h^{-1} + \log \log h^{-1} - 2 \log \log \log h^{-1} - \log(\pi-\varepsilon))^{1/2}. \tag{3}$$

That is to say

$$a_3(h)\in LUC(c_1)$$

for any $\varepsilon > 0$.

Theorem 2. For any $\varepsilon > 0$ and for almost all $\omega\in\Omega$ there exists a $0 < h_0 = h_0(\omega,\varepsilon) < 1$ such that

$$c_1(h) \ge a_4(h) \tag{4}$$

if $h \le h_0$ where

$$a_4(h) = a_4(h,\omega) = (2 \log h^{-1} + \log\log h^{-1} - 2 \log\log\log h^{-1} - \log(9\pi+\varepsilon))^{1/2} . \qquad (5)$$

That is to say

$$a_4(h) \in LLC(c_1)$$

for any $\varepsilon > 0$.

Let us formualte a trivial consequence of Theorems B and 1, giving a result much stronger than Theorem A (at least in the case $i = 1$):

Theorem 3. We have

$$\lim_{h \to 0} (c_1(h) - (2 \log h^{-1})^{1/2}) = 0 \quad \text{a.s.}$$

The author is indepted to Professor H. Rootzen for his valuable remarks.

II. SOME LEMMAS

Our first three lemmas are well-known:

Lemma A. (Qualls-Watanabe, 1972) For any fixed $k > 0$ we have

$$P\{\sup_{0 \le x \le k} (W(x+1)-W(x)) > u\} = (1 + o(1)) \frac{k}{\sqrt{2\pi}} u e^{-\frac{u^2}{2}}$$

as $u \to \infty$.

Lemma B. (Slepian, 1962). Let $G(t)$ and $G^*(t)$ be Gaussian processes (possessing continuous sample functions). Suppose that these are standardized so that $EG(t) = EG^*(t) = 0$, $EG^2(t) = E(G^*(t))^2 = 1$ and write $\rho(t,s)$ and $\rho^*(t,s)$ for their covariance functions. Suppose that $\rho(t,s) \geq \rho^*(t,s)$. Then

$$P\{\sup_{0\leq t\leq T} |G(t)| \leq u\} \geq P\{\sup_{0\leq t\leq T} |G^*(t)| \leq u\}.$$

Lemma C. (Erdös-Rényi, 1959). If $A_1, A_2, \ldots$ are arbitrary events, fulfilling the conditions

$$\sum_{n=1}^{\infty} P(A_n) = \infty$$

and

$$\liminf_{n\to\infty} \frac{\sum_{k=1}^{n} \sum_{\ell=1}^{n} P(A_k A_\ell)}{\left(\sum_{k=1}^{n} P(A_k)\right)^2} = 1$$

there there occur with probability 1 infinitely many of the events A_n.

Lemma 1. For any $\varepsilon > 0$ there exist a $u_0 = u_0(\varepsilon) > 0$ and a $T_0 = T_0(\varepsilon) > 0$ such that

$$\exp\{-(1+\varepsilon)\frac{T}{\sqrt{2\pi}}ue^{-\frac{u^2}{2}}\} \leq P\{\sup_{0\leq x\leq T}(W(x+1)-W(x)) \leq u)$$
$$\leq \exp\{-(1-\varepsilon)\frac{T}{\sqrt{2\pi}}ue^{-\frac{u^2}{2}}\}$$

if $u \geq u_0$ and $T \geq T_0$.

Proof. Let

$$X_i = \sup_{i\leq x\leq i+1}(W(x+1)-W(x)) \qquad (i = 0,1,2,\ldots),$$

$$Y(a,b) = \max_{a\leq i\leq b} X_i = \sup_{a\leq x\leq b+1}(W(x+1)-W(x))$$

where a and b are positive integers. Then by Lemmas A and B for any $T = 1,2,\ldots$ we have

$$P\{\sup_{0\leq x\leq T}(W(x+1)-W(x)) \leq u\} = P\{\max_{0\leq i\leq T-1} X_i \leq u\} \geq (P(X_1 \leq u))^T$$
$$= (1-(1+o(1))\frac{u}{\sqrt{2\pi}}e^{-\frac{u^2}{2}})^T = \exp(-(1+o(1))\frac{u}{\sqrt{2\pi}}Te^{-\frac{u^2}{2}}).$$

On the other hand if $k < T$ are positive integers then

$$\sup_{0\leq x\leq T}(W(x+1)-W(x)) = \max(Y(0,k-1),\ X_k,\ Y(k+1,\ 2k), X_{2k+1},\ldots,$$
$$Y(\ell(k+1),(\ell+1)(k+1)-2, Y((\ell+1)(k+1)-1,T-1))$$
$$\geq \max(Y(0,k-1), Y(k+1,2k),\ldots,Y(\ell(k+1),$$
$$(\ell+1)(k+1)-2))$$

where ℓ is the largest integer for which $(\ell+1)(k+1)-1 \le T$. Hence by Lemma A we have

$$P\{\sup_{0\le x\le T}(W(x+1)-W(x))\le u\} \le (P\{\sup_{0\le x\le k}(W(x+1)-W(x))\le u\})^{\ell+1}$$

$$= (1-(1+o(1))\frac{k}{\sqrt{2\pi}}ue^{-\frac{u^2}{2}})^{\ell+1} = \exp\{-(1+o(1))\frac{k(\ell+1)}{\sqrt{2\pi}}ue^{-\frac{u^2}{2}}\}$$

which proves Lemma 1.

A trivial consequence of the above Lemma is:

Lemma 2. For any $\varepsilon > 0$ there exist a $0 < h_0 = h_0(\varepsilon) < 1$ and a $u_0 = u_0(\varepsilon) > 0$ such that

$$\exp\{-\frac{1+\varepsilon}{h\sqrt{2\pi}}ue^{-\frac{u^2}{2}}\} \le P\{\sup_{0\le x\le 1-h} h^{-1/2}(W(x+h)-W(x)) \le u\}$$

$$\le \exp\{-\frac{1-\varepsilon}{h\sqrt{2\pi}}ue^{-\frac{u^2}{2}}\}$$

if $u \ge u_0$ and $h \le h_0$.

III. THE PROOF OF THEOREM 2.

Apply Lemma 2 with $u = a_4(h,\varepsilon_1) = a_4(h)$ of (5). Then we have

$$P\{c_1(h) \le a_4(h)\} \le C(\log h^{-1})^{-3-\delta} \tag{6}$$

with some $\delta > 0$. Replacing the h by $h_n = \exp\{-n^{1/3-\rho}\}$ $(0 < \rho < 1/3)$, if ρ is small enough we got that among the events $c_1(h_n) \le a_4(h_n)$ only finitely many will occur with probability 1.

Let $\tau_n = n^{-(\xi+2/3)} \exp\{-n^{1/3-\xi}\} \geq h_{n-1} - h_n$ and $h_n \leq h \leq h_{n-1}$. Then by Theorem A we have

$$\sup_{0\leq t\leq 1-\tau_n} \sup_{0\leq s\leq \tau_n} |W(t+s)-W(t)| \leq \left((2+\varepsilon)\tau_n \log \tau_n^{-1}\right)^{1/2}$$

with probability 1 for any $\varepsilon > 0$ if n is large enough. This implies

$$c_1(h) \geq a_4(h_n) - \left((2+\varepsilon)\tau_n \log \tau_n^{-1}\right)^{1/2} \geq a_4(h,2\varepsilon)$$

which proves (4).

IV. THE PROOF OF THEOREM 1

Apply Lemma 2 with $u = a_3(h,\varepsilon) = a_3(h)$. Then we have

$$P\{\sup_{0\leq x\leq 1-h} h^{-1/2}(W(x+h)-W(x)) \leq a_3(h)\} \leq (\log h^{-1})^{-1+\delta} \tag{7}$$

with some $\delta > 0$. Choosing $h_n = \exp\{-n^{1+\rho}\}$ we get

$$\sum_{n=1}^{\infty} P\{c_1(h_n) \leq a_3(h_n)\} = \infty \tag{8}$$

if ρ is small enough. Since the events $A_n = \{c_1(h_n) \leq a_3(h_n)\}$ are non-independent the original form of the Borel-Cantelli lemma cannot be applied. We show that Lemma C can be applied in this situation. At first we present two lemmas.

<u>Lemma 3</u>. For any $\varepsilon > 0$ there exists an integer $n_0 = n_0(\varepsilon)$ such that

$$P(A_{n+i}|\overline{A}_n) \geq (1 - \varepsilon P(A_n))P(A_{n+i}) \tag{9}$$

if $n \geq n_0$ and $i = 1,2,\ldots$.

Proof. For any integer n we have

$$\bar{A}_n = \bigcup_{0\le\xi\le 1-h_n} \bigcup_{a_3(h_n)\le z\le(4\log h_n^{-1})^{1/2}} \{h_n^{-1/2}(W(\xi+h_n)-W(\xi))=z\}$$

$$+ \{\sup_{0\le x\le 1-h_n} h_n^{-1/2}(W(x+h_n)-W(x)) \ge (4\log h_n^{-1})^{1/2}\} = B_n^{(1)} + B_n^{(2)}.$$

An easy calculation by Lemma 2 gives

$$P(B_n^{(2)}) \le \frac{\varepsilon}{3} P(A_n)P(B_n^{(1)}) \tag{10}$$

if n is large enough.

Let now $z \le (4\log h_n^{-1})^{1/2}$ then we have

$$P(A_{n+i}|h_n^{-1/2}(W(\xi+h_n)-W(\xi)) = z\} \ge P\{\sup_{0\le x\le 1-h_{n+i}} (W(x+h_{n+i})-W(x))$$
$$\le h_{n+i}^{1/2} a_3(h_{n+i}) - h_n^{1/2}(4\log h_n^{-1})^{1/2} \frac{h_{n+i}}{h_n}\} \ge (1-(\varepsilon/3)P(A_n))P(A_{n+i}). \tag{11}$$

(11) implies:

$$P(A_{n+i}|B_n^{(1)}) \ge (1-(\varepsilon/3)P(A_n))P(A_{n+i}) . \tag{12}$$

By (10) and (12) we have

$$P(A_{n+i}|\bar{A}_n) = P(A_{n+i}|B_n^{(1)} + B_n^{(2)}) = \frac{P(A_{n+i}B_n^{(1)} + A_{n+i}B_n^{(2)})}{P(B_n^{(1)}) + P(B_n^{(2)})}$$

$$\ge P(A_{n+i}|B_n^{(1)}) \frac{P(B_n^{(1)})}{P(B_n^{(1)}) + P(B_n^{(2)})} \ge \frac{1-(\varepsilon/3)P(A_n)}{1+(\varepsilon/3)P(A_n)} P(A_{n+i})$$

$$\ge (1-\varepsilon P(A_n))P(A_{n+i})$$

hence (9) is proved.

<u>Lemma 4.</u> For any $\varepsilon > 0$ there exists an integer $n_0 = n_0(\varepsilon)$ such that

$$P(A_n A_{n+i}) \leq (1+\varepsilon)P(A_n)P(A_{n+i}) \tag{13}$$

if $n \geq n_0$ and $i = 1,2,\ldots$.

Proof. By Lemma 3 we have

$$P(A_{n+i}A_n) = P(A_{n+i})-P(A_{n+i}\overline{A}_n) = P(A_{n+i})-P(A_{n+i}|\overline{A}_n)P(\overline{A}_n)$$

$$\leq P(A_{n+i})-(1-\varepsilon P(A_n))P(A_{n+i})P(\overline{A}_n)$$

$$= P(A_{n+i})[1-(1-\varepsilon P(A_n))(1-P(A_n))]=(1+\varepsilon-\varepsilon P^2(A_n))P(A_n)P(A_{n+i})$$

which proves (13).

Now, Theorem 1 follows from (8), (13) and Lemma C.

REFERENCES

[1] Chung, K.L., Erdös, P., Sirao, T. (1959). On the Lipschitz condition for Brownian motion. <u>J. of Math. Soc. Japan</u>, <u>11</u>, p. 263-274.

[2] Erdös, P., Rényi, A. (1959). On Cantor's series with convergent $\Sigma\, 1/q_n$. <u>Ann. Univ. Sci. Budapest</u>, R. Eötvös nom. Sect. Math., <u>2</u>, p. 93-109.

[3] Erdös, P., Révész, P. (1976). On the length of the longest head-run. Coll. Math. Soc. J. Bolyai, 16 Topics in Information Theory, Keszthely, Hungary 1975, North Holland, Amsterdam.

[4] Lévy, P. (1937). Théorie de l'addition des variables aléatoires. Gauthier-Villars, Paris.

[5] Pruitt, W.E., Orey, S. (1973). Sample functions of the N-parameter Wiener processes. Ann. Probability, 1, p. 138-163.

[6] Slepian, D. (1962). The one-sided barrier problem for Gaussian noise. Bell. Syst. Tech. J., 41, p. 463-501.

[7] Quall, C., Watanabe, H. (1972). Asymptotic Properties of Gaussian Processes. Ann. Math. Statist., 43, p. 580-596.

ASYMPTOTIC SEPARATION OF DISTRIBUTIONS AND CONVERGENCE PROPERTIES OF TESTS AND ESTIMATORS

Jean Geffroy

Institut de Statistique
Université Pierre et Marie Curie
Paris, France

The aim of this paper is to introduce the concept of "decantation of distributions" as an efficient tool to prove asymptotic separation properties of two families of distribution sequences, in the frame of fairly general statistical models. A fundamental theorem is given, and its application to a problem of consistent estimation is worked out.

I.

In this paper, we are concerned with estimators and tests founded on a sequence (X_n) of random variables not necessarily independent or equidistributed. Each variable X_n takes its value in a measurable space $(\mathcal{X}_n, \mathcal{B}_n)$. We put: $(\mathcal{X}^{(n)}, \mathcal{B}^{(n)}) = \otimes_{i=1}^{n} (\mathcal{X}_i, \mathcal{B}_i)$ $(n = 1,2,\ldots,\infty)$.

Our fundamental hypothesis is that the joint probability distribution (p.d.) of the sequence $(X_1, X_2, \ldots, X_n, \ldots)$ belongs to a given family $\{P_\theta\}_{\theta\in\Theta}$ of p.d. on $(\mathcal{X}^{(\infty)}, \mathcal{B}^{(\infty)})$. Sometimes, we write $P_\theta^{(\infty)}$ instead of P_θ, and similarly we represent by $P_\theta^{(n)}$ the projection of P_θ on $(\mathcal{X}^{(n)}, \mathcal{B}^{(n)})$, i.e. the joint p.d. of $X^{(n)} = (X_1, \ldots, X_n)$ in the hypothesis θ. The projection of P_θ on $\mathcal{X}_n$ is $P_{n,\theta}$.

It will be convenient to suppose that P_θ is defined by $P_{1,\theta}$ and a sequence of probability transitions $P_{n,\theta}^{x^{(n-1)}}$ $(n \geq 2)$ where $x^{(n-1)}$ runs in $\mathcal{X}^{(n-1)}$. We shall use also the notation $P_{n,\theta}(\cdot/x^{(n-1)})$ for

ISBN: 0-12-166650-6

these p.t. In order to avoid fussy distinctions, we adopt the conventions:

$$P_{1,\theta}^{x^{(0)}} = P_{1,\theta} \quad \text{and} \quad \mathcal{X}^{(0)} = \{x^{(0)}\}.$$

II.

Definition. Let Θ_0 and Θ_1 be two disjoint non-void parts of Θ. We say the two families of sequences $\{P_\theta^{(n)}\}_{\theta\in\Theta_0}$ and $\{P_\theta^{(n)}\}_{\theta\in\Theta_1}$ are asymptotically separated if there exists a sequence of sets $D_n \in \mathcal{B}^{(n)}$ satisfying:

$$(\forall\theta \in \Theta_0) \lim_{n\to\infty} P_\theta^{(n)}(D_n) = 1 \tag{1}$$

$$(\forall\theta \in \Theta_1) \lim_{n\to\infty} P_\theta^{(n)}(D_n) = 0. \tag{2}$$

When Θ_0 and Θ_1 are complementary sets, i.e. $\Theta_0 \cup \Theta_1 = \Theta$, the above property is equivalent to the existence of a convergent sequence of tests $T_n(X_1,\ldots,X_n)$ of the hypothesis $\{\theta \in \Theta_0\}$ against $\{\theta \in \Theta_1\}$. Moreover, one knows that, if the asymptotic separation of $\{P_\theta^{(n)}\}_{\Theta_0}$ and $\{P_\theta^{(n)}\}_{\Theta_1}$ stands whenever Θ_0 and Θ_1 are respectively "well chosen" neighborhoods of two arbitrary points θ_0 and θ_1 for some topology on Θ, that often implies the existence of a consistent estimator of θ (cf §7). So investigations about asymptotic separation of certain families of distributions may constitute the first step in studying convergence properties of tests or estimators. A characteristic feature of our own approach is the systematic use of a new concept that we are calling "decantation."

III.

Definition.

a) Let $\{P_\theta\}_{\theta\in\Theta_0}$ and $\{P_\theta\}_{\theta\in\Theta_1}$ be two families of probability measures on a measurable space (Ω,A) and B a set of A. We say that B is decanting the first family from the second if we have:

$$\sup_{\theta\in\Theta_0} P_\Theta(B) < \inf_{\theta\in\Theta_1} P_\theta(B). \qquad (3)$$

b) For each $n = 1,2,\ldots$ let $\{P_{n,\sigma_n,\theta}\}^{\sigma_n\in\Sigma_n}_{\theta\in\Theta_0}$ and $\{P_{n,\sigma_n,\theta}\}^{\sigma_n\in\Sigma_n}_{\theta\in\Theta_1}$ be two families of probability measures on a measurable space (Ω_n,A_n) and $\{B_{n,\sigma_n}\}_{\sigma_n\in\Sigma_n}$ a family of sets of A_n. We say the first p.d. family is asymptotically decanted from the second one by the sets $\{B_{n,\sigma_n}\}_{\sigma_n\in\Sigma_n}$ if there exists two monotonous nondecreasing sequences $\Theta_{i,n} \subset \Theta_i$ $(i = 0,1)$ converging respectively towards Θ_0 and Θ_1 when $n \to \infty$, and such that, for each n:

$$\sup_{\theta\in\Theta_{0;n};\sigma_n\in\Sigma_n} P_{n,\sigma_n,\theta}(B_{n,\sigma_n}) < \inf_{\theta\in\Theta_{1,n};\sigma_n\in\Sigma_n} P_{n,\sigma_n,\theta}(B_{n,\sigma_n}). \qquad (4)$$

(We give the value 0 to the first member if $\theta_{0,n}$ is empty, and the value 1 to the second member if $\Theta_{1,n}$ is empty.)

In the sequel, we shall use the following notations:

$$(\forall\theta \in \Theta_0)\ N_0(\theta) = \max\{j/(j=0) \cup (\theta \notin \Theta_{0,j})\}$$

$$(\forall\theta \in \Theta_1)\ N_1(\theta) = \max\{j/(j=0) \cup (\theta \notin \Theta_{1,j})\}.$$

Remark. If (3) or (4) holds with $\leq$ instead of $<$, we say there is decantation in the weak sense.

IV.

Coming back to our model, we are now considering two families $\{P_\theta\}_{\Theta_0}$ and $\{P_\theta\}_{\Theta_1}$ of p.d. on $(\mathcal{X}^{(\infty)}, \mathcal{B}^{(\infty)})$. For each $n \geq 1$, let B_n be an application of $\mathcal{X}^{(n-1)}$ into $\mathcal{B}^{(n)}$:

$$x^{(n-1)} \to B_n(x^{(n-1)})$$

Of course, the function $B_1(\cdot)$ has a unique value also noted B_1. The characteristic function $1_{B_n(x^{(n-1)})}(x_n)$ is, in fact, a function of $x^{(n)}$. We shall write it $\xi_n(x^{(n)})$. As a regularity hypothesis, we suppose that ξ_n is measurable, which implies that, for every $\theta \in \Theta$, $\xi_n(X^{(n)})$ is a random variable.

The probability transitions $P_{n,\theta}(\cdot/x^{(n-1)})$ induce a family of conditional p.t. $P_{n,\theta}(\cdot/\xi^{(n-1)})$, where $\xi^{(n-1)} = (\xi_1,\ldots,\xi_{n-1})$. In the particular case $n = 1$, the symbol $P_{1,\theta}(\cdot/\xi^{(0)})$ means only $P_{1,\theta}$.

Our decantation assumptions are concerning these conditional p.d. We suppose that, for some choice of B_n $(n = 1,2,\ldots)$ there exists $\Theta_{0,n}\uparrow\Theta_0$ and $\Theta_{1,n}\uparrow\Theta_1$ giving rise, for each n, to:

$$\bar{\omega}_{0,n} \leq \bar{\omega}_{1,n} \tag{5}$$

where $\bar{\omega}_{0,n}$ and $\bar{\omega}_{1,n}$ are defined by:

$$\bar{\omega}_{0,n} = \sup_{\theta\in\Theta_{0,n};x^{(n-1)}\in\mathcal{X}^{(n-1)}} P_{n,\theta}[B_n(x^{(n-1)})/\xi^{(n-1)}(x^{(n-1)})] \tag{5'}$$

$$\bar{\omega}_{1,n} = \inf_{\theta\in\Theta_{1,n};x^{(n-1)}\in\mathcal{X}^{(n-1)}} P_{n,\theta}[B_n(x^{(n-1)})/\xi^{(n-1)}(x^{(n-1)})]. \tag{5''}$$

We can then state the following result.

V.

Theorem 1. Under the decantation hypothesis assumed in §4, let us put $\alpha_n = \overline{\omega}_{1,n} - \overline{\omega}_{0,n}$ (n = 1,2,...) and define:

$$Y_n = \sum_{i=1}^{n} \alpha_i \xi_i (X^{(i)}).$$

Then, we have:

a) for every $\theta \epsilon \Theta_0$ and $n > N_0(\theta)$

$$P_\theta^{(n)}\{Y_n > \frac{1}{2} \sum_1^n (\overline{\omega}_{1,i}^2 - \overline{\omega}_{0,i}^2)\} \leq \exp\left[\sum_i^{N_0(\theta)} \alpha_i - \frac{1}{4} \sum_1^n \alpha_i^2\right], \quad (6)$$

b) for every $\theta \epsilon \Theta_1$ and $n > N_1(\theta)$

$$P_\theta^{(n)}\{Y_n < \frac{1}{2} \sum_1^n (\overline{\omega}_{1,i}^2 - \overline{\omega}_{0,i}^2)\} \leq \exp\left[\sum_1^{N_1(\theta)} \alpha_i - \frac{1}{4} \sum_1^n \alpha_i^2\right]. \quad (7)$$

($N_0(\theta)$ and $N_1(\theta)$ are defined as in §3.)

Proof. Be definition of ξ_n we have, for each $n \geq 1$:

$$P_{n,\theta}[B_n(x^{(n-1)})/\xi^{(n-1)}(x^{(n-1)})] = E_\theta[\xi_n(X^{(n)})/\xi^{(n-1)}(x^{(n-1)})].$$

(The conditioning is symbolic for $n = 1$.) Supposing $\theta \epsilon \Theta_0$ and $n > N_0(\theta)$, formula (5') yields:

$$(\forall x^{(n-1)} \epsilon \mathcal{X}^{(n-1)}),\ E_\theta[\xi_n(X^{(n)})/\xi^{(n-1)}(x^{(n-1)})] \leq \overline{\omega}_{0,n}. \quad (8)$$

Similarly, for $\theta\in\Theta_1$ and $n > N_1(\theta)$, formula (5") yields:

$$(\forall x^{(n-1)})\in X^{(n-1)},\ E_\theta[\xi_n(X^{(n)})/\xi^{(n-1)}(x^{(n-1)})] \geq \bar{\omega}_{1,n}. \tag{9}$$

Putting $n_0 = N_0(\theta)$ if $\theta\in\Theta_0$ and $n_1 = N_1(\theta)$ if $\theta\in\Theta_1$, we see that the random variables $\xi_1,\ldots,\xi_n$ obtained for the corresponding values of θ satisfy respectively the conditions (i) and (ii) of the following lemma. This last result being admitted (cf. [2]), theorem 1 is proved by choosing $\gamma_j = \alpha_j$.

<u>Lemma 1.</u> Let ξ_n $(n = 1,2,\ldots)$ be a sequence of simultaneous Bernoulli variates, γ_n non-random numbers in $[0,1]$, and $Y_n = \Sigma_1^n \gamma_j\xi_j$.

i) $\bar{\omega}_{0,n}$ $(n = 1,2,\ldots)$ being a number of $[0,1]$ and n_0 a non-negative integer, suppose the conditions:

$$E(\xi_1) \leq \bar{\omega}_{0,1};\ E(\xi_2/\xi_1) \leq \bar{\omega}_{0,2};\ldots;\ E(\xi_n/\xi_1,\ldots,\xi_{n-1}) \leq \bar{\omega}_{0,n};\ldots$$

hold whenever $n > n_0$. For these values of n, we have:

$$\Pr\{Y_n > \sum_1^n \gamma_j(\bar{\omega}_{0,j} + \tfrac{1}{2}\gamma_j)\} \leq \exp(\sum_1^{n_0} \gamma_j - \tfrac{1}{4}\sum_1^n \gamma_j^2).$$

ii) $\bar{\omega}_{1,n}$ $(n = 1,2,\ldots)$ being a number of $[0,1]$ and n_1 a non-negative integer, suppose the conditions:

$$E(\xi_1) \geq \bar{\omega}_{1,1};\ E(\xi_2/\xi_1) \geq \bar{\omega}_{1,2};\ldots;\ E(\xi_n/\xi_1,\ldots,\xi_{n-1}) \geq \bar{\omega}_{1,n}$$

hold whenever $n > n_1$. For these values of n, we have:

$$\Pr\{Y_n \leq \sum_1^n \gamma_j(\bar{\omega}_{1,j} - \tfrac{1}{2}\gamma_j)\} \leq \exp(\sum_1^{n_1} \gamma_j - \tfrac{1}{4}\sum_1^n \gamma_j^2).$$

Remark. The coefficient $1/4$ before $\sum \alpha_i^2$ is not at all "optimal." It is possible to replace it by any $c < 1/2$, but we do not know if one can put $c = 1/2$.

The next result is an immediate but useful consequence of theorem 1.

Theorem 2. With the same hypotheses as in theorem 1, there exists a sequence of sets $D_n \in \mathcal{B}^{(n)}$ satisfying the two conditions hereafter:

a) for every $\theta \in \Theta_0$ and $n > N_0(\theta)$

$$P_\theta^{(n)}(D_n) \geq 1 - \exp\left[\sum_1^{N_0(\theta)} \alpha_i - \frac{1}{4}\sum_1^n \alpha_i^2\right],$$

b) for every $\theta \in \Theta_1$ and $n > N_1(\theta)$

$$P_\theta^{(n)}(D_n) \leq \exp\left[\sum_1^{N_1(\theta)} \alpha_i - \frac{1}{4}\sum_1^n \alpha_i^2\right].$$

VI.

When the r.v.'s X_n are independent, the p.t. $P_{n,\theta}^{x^{(n-1)}}$ become identical to the marginal distributions $P_{n,\theta}$. In this case, there is no need to adapt the set B_n to the r.v. $x^{(n-1)}$ and consequently B_n is chosen as a fixed set in $\mathcal{B}_n$, independent of $x^{(n-1)}$.

Let us consider, for a while, the subcase in which Θ_0 and Θ_1 contain just one element: $\Theta_0 = \{0\}$, $\Theta_1 = \{1\}$. Then, it seems interesting to choose each B_n maximizing $|P_{n,1}(B) - P_{n,0}(B)|$ $(B \in \mathcal{B}_n)$. As is well known, this is possible, and we have: $P_{n,1}(B_n) - P_{n,0}(B_n) = d_V(P_{n,1}, P_{n,0})$ where d_V represents the variational distance on any set M_n of p.d. But in this case, $\overline{\omega}_{0,n} = P_{n,0}(B_n)$ and $\overline{\omega}_{1,n} = P_{n,1}(B_n)$ hold, and so $\alpha_n = d_V(P_{0,n}, P_{1,n})$.

Theorem 2 clearly implies

$$d_V(P_0^{(n)}, P_1^{(n)}) \geq 1 - 2\exp\left(-\frac{1}{4}\sum_1^n \alpha_i^2\right). \tag{10}$$

By this relation, the condition:

$$\sum_1^\infty \alpha_i^2 = \infty \tag{11}$$

is sufficient for the asymptotic separation of $P_0^{(n)}$ and $P_1^{(n)}$.

Conversely, W. Hoeffding and J. Wolfowitz established the inequality:

$$d_V(P_0^{(n)}, P_1^{(n)}) \leq 1 - \prod_{i=1}^{n} (1-\alpha_i). \tag{12}$$

If there exists i_0 such that $\alpha_{i_0} = 1$, then $P_0^{(n)}$ and $P_1^{(n)}$ are separate whenever $n \geq i_0$. In the general case, $\alpha_n \neq 1$ stands for all n, and then, by (12), a necessary condition for asymptotic separation of $P_0^{(n)}$ and $P_1^{(n)}$ is:

$$\sum_1^\infty \alpha_i = \infty. \tag{13}$$

Although it is possible to prove inequalities more stringent than (10) (cf. [10] and [12]), conditions (11) and (13) possess a sort of optimality. A. Hillion (cf. [5]) has shown that there is no sufficient condition strictly weaker than (11), and no necessary condition strictly stronger than (13). An interesting consequence is the nonexistence of a necessary and sufficient condition for the asymptotic separation of $P_0^{(n)}$ and $P_1^{(n)}$ which would be found only on the sequence $d_V(P_{n,0},P_{n,1})$.

VII.

In order to illustrate the use of theorem 1, we are going to deal with a problem generalizing the translation parameter estimation problem.

The statistical model is defined as follows.

- All the X_n are real random variables.
- The index set Θ is a bounded interval of $\mathbb{R}$.
- P is a given continuous probability measure on $\mathbb{R}$.
- Each p.t. $P^{x^{(n-1)}}_{n,\theta}$ $(n \geq 1)$ is deduced from P by the translation $x \to x + \lambda_n(x^{(n-1)}, \theta)$, where λ_n is, for any fixed $x^{(n-1)}$, a monotonous non-decreasing real function on Θ.
- The functions λ_n satisfy the condition:

$$(\forall n \geq 1), (\forall x^{(n-1)} \in \mathbb{R}^{n-1}), (\forall\theta\in\Theta, \forall\theta'\in\Theta, \theta' < \theta)$$
$$\underline{h}(\theta-\theta') \leq \lambda_n(x^{(n-1)}, \theta) - \lambda_n(x^{(n-1)}, \theta') \leq \overline{h}(\theta-\theta') \tag{14}$$

where $\underline{h}$ and $\overline{h}$ are positive continuous real functions, monotonous increasing and such that

$$\lim_{\delta\to 0+} \underline{h}(\delta) = \lim_{\delta\to 0+} \overline{h}(\delta) = 0. \tag{15}$$

Our aim is to establish the existence of a consistent estimator of θ. More precisely, we shall prove

<u>Theorem 3</u>. In the statistical model stated above, the parameter θ has an estimator $\hat{\theta}_n(X_1,\ldots,X_n)$ converging in probability uniformly. This means that, for every $\varepsilon > 0$, there exists $n_0 \in \mathbb{N}$ giving rise to

$$(\forall\theta\in\Theta),(\forall n>n_0)P^{(n)}_\theta(|\hat{\theta}_n-\theta| > \varepsilon) < \varepsilon. \tag{a}$$

Furthermore, it is possible to construct $\hat{\theta}_n$ in such a way that it is converging almost completely surely uniformly, which means that to every $\varepsilon > 0$, one can associate $n_0 \in \mathbb{N}$ giving rise to:

$$(\forall\theta\in\Theta), \quad \sum_{n=n_0}^{\infty} P_\theta^{(n)}(|\hat{\theta}_n - \theta| > \varepsilon) < \varepsilon . \qquad \text{(b)}$$

VIII.

Proof. It will be convenient to use the symbol $P_{[\lambda]}$ to label the p.d. deduced from P by the translation $x \to x + \lambda$. Similarly, F, being the distribution function (d.f.) of P, $F_{[\lambda]}$ designates the d.f. of $P_{[\lambda]}$. We have the identity

$$F_{[\lambda]}(x) = F(x-\lambda).$$

If P_0 and P_1 are two arbitrary p.d. on $\mathbb{R}$, F_0 and F_1 their respective d.f., the symbol $d_K(P_0,P_1)$ or $d_K(F_0,F_1)$ represents the "Kolmogorov's distance" of P_0 and P_1:

$$d_K(P_0,P_1) = d_K(F_0,F_1) = \sup_{(x)} | F_0(x) - F_1(x)| .$$

For any value of λ we have

$$d_K(P; P_{[\lambda]}) = \sup_{(x)} |F(x)-F_{[\lambda]}(x)| = \max_{(x)} [F(x) - F(x-\lambda)].$$

But we recognize, in the last expression, the concentration function $Q(\lambda)$ of the p.d. P, and therefore:

$$d_K(P, P_{[\lambda]}) = Q(\lambda).$$

We now prove an auxiliary result.

Lemma 2. Let θ and θ' be arbitrary points in Θ with, for instance, $\theta' < \theta$. Then we have, for every $n \geq 1$ and $x^{(n-1)} \in X^{(n-1)}$:

$$Q[\underline{h}(\theta-\theta')] \leq d_K(P^{x^{(n-1)}}_{n,\theta}, P^{x^{(n-1)}}_{n,\theta'}) \leq Q[\overline{h}(\theta-\theta')]. \qquad (16)$$

* To see that, we note the relation:

$$(\forall\theta\in\Theta),\ P^{x^{(n-1)}}_{n,\theta} = P_{[\lambda_n(x^{(n-1)},\theta)]}$$

which yields

$$d_K(P^{x^{(n-1)}}_{n,\theta}, P^{x^{(n-1)}}_{n,\theta'}) = d_K(P_{[\lambda_n(x^{(n-1)},\theta)]}, P_{[\lambda_n(x^{(n-1)},\theta')]})$$

$$= d_K(P, P_{[\lambda_n(x^{(n-1)},\theta) - \lambda_n(x^{(n-1)},\theta')]})$$

$$= Q[\lambda_n(x^{(n-1)},\theta) - \lambda_n(x^{(n-1)},\theta')].$$

Every concentration function being monotonous non-decreasing, the lemma results immediately from (14).

IX.

Let θ_0 and θ_1 $(\theta_1 < \theta_0)$ be two arbitrary distinct values in Θ, and δ' a number satisfying $0 < \delta' \leq \theta_0 - \theta_1$. We can choose a number $\overline{x} = \overline{x}(\theta_0 - \theta_1)$ such that

$$F(\overline{x}) - F[\overline{x} - \underline{h}(\theta_0-\theta_1)] = Q[\underline{h}(\theta_0-\theta_1)] \geq Q[\underline{h}(\delta')] \qquad (17)$$

Besides, $F^{x^{(n-1)}}_{n,\theta}$ being the d.f. of $P^{x^{(n-1)}}_{n,\theta}$, we have clearly, for every n (≥ 1) and $x^{(n-1)} \in \mathbb{R}^{n-1}$,

$$F^{x^{(n-1)}}_{n,\theta_1}[\overline{x} + \lambda_n(x^{(n-1)},\theta_1)] = F(\overline{x})$$

and

$$F^{x^{(n-1)}}_{n,\theta_0}[\overline{x} + \lambda_n(x^{(n-1)},\theta_1)] = F\{\overline{x} - [\lambda_n(x^{(n-1)},\theta_0) - \lambda_n(x^{(n-1)},\theta_1)]\}$$

$$\leq F(\overline{x} - \underline{h}(\theta_0-\theta_1)].$$

Taking

$$B_n(x^{(n-1)}) =]-\infty, \overline{x} + \lambda_n(x^{(n-1)},\theta_1)]$$

the preceding relations give us, for all possible n and $x^{(n-1)}$:

$$P_{n,\theta_1}[B_n(x^{(n-1)})/x^{(n-1)}] = F(\overline{x}) \tag{18}$$

$$P_{n,\theta_0}[B_n(x^{(n-1)})/x^{(n-1)}] \leq F[\overline{x} - \underline{h}(\theta_0-\theta_1)]. \tag{19}$$

X.

Our next step is to compare the families $\{P^{x^{(n-1)}}_{n,\theta}\}$ with $|\theta-\theta_0| < \delta$ or $|\theta-\theta_1| < \delta$, where δ is, at the beginning, an arbitrary positive number.

We deduce from the inequalities (16):

$$(\forall\theta\in\Theta,\ |\theta-\theta_0| < \delta),\ (\forall x^{(n-1)} \in \mathbb{R}^{n-1})$$

$$d_K(P^{x^{(n-1)}}_{n,\theta}, P^{x^{(n-1)}}_{n,\theta_0}) \leq Q[\overline{h}(\theta-\theta_0|)] \leq Q[\overline{h}(\delta)].$$

Taking into account the inequality (19) and the fact that $B_n(x^{(n-1)})$ is a half-line, the above relation implies:

$$(\forall\theta\in\Theta,\ |\theta-\theta_0| < \delta),\ (\forall x^{(n-1)} \in \mathbb{R}^{n-1}),$$
$$P_{n,\theta}[B_n(x^{(n-1)})/x^{(n-1)}] \le F[\overline{x} - \underline{h}(\theta_0-\theta_1)] + Q[\overline{h}(\delta)]. \tag{20}$$

An analogous calculation, based upon (16) and (18), leads to:

$$(\forall\theta\in\Theta,\ |\theta-\theta_1| < \delta),\ (\forall x^{(n-1)} \in \mathbb{R}^{n-1})$$
$$P_{n,\theta}[B_n(x^{(n-1)})/x^{(n-1)}] \ge F(\overline{x}) - Q[\overline{h}(\delta)]. \tag{21}$$

Let us put:

$$\overline{\omega}_{0,n} = F[\overline{x} - \underline{h}(\theta_0-\theta_1)] + Q[\overline{h}(\delta)]$$

and

$$\overline{\omega}_{1,n} = F(\overline{x}) - Q[\overline{h}(\delta)].$$

We have:

$$\overline{\omega}_{1,n} - \overline{\omega}_{0,n} = F(\overline{x}) - F[\overline{x} - \underline{h}(\theta_0-\theta_1)] - 2Q[\overline{h}(\delta)]$$
$$\ge Q[\underline{h}(\delta')] - 2Q[\overline{h}(\delta)].$$

By the continuity properties of $\underline{h}$, $\overline{h}$ and Q, if δ' is small enough (say $\delta' \le \delta'_0$), we can choose δ such that:

$$Q[\underline{h}(\delta')] = 3Q[\overline{h}(\delta)]. \tag{22}$$

This gives us $\overline{\omega}_{1,n} - \overline{\omega}_{0,n} \ge Q[\overline{h}(\delta)]$. Observing that (20) and (21) imply respectively, for all n and $x^{(n-1)}$:

$$(\forall\theta\in\Theta,\ |\theta-\theta_0| < \delta),\ P_{n,\theta}[B_n(x^{(n-1)})/\xi^{(n-1)}] \le \bar{\omega}_{0,n}$$

and

$$(\forall\theta\in\Theta,\ |\theta-\theta_1| < \delta),\ P_{n,\theta}[B_n(x^{(n-1)})/\xi^{(n-1)}] \ge \bar{\omega}_{1,n}$$

we see that (5) is satisfied if we take

$$\Theta_j = \Theta_{j,n} = \{\theta\in\Theta/|\theta-\theta_j| < \delta\} \quad (j = 0,1).$$

Applying theorem 2, we obtain

Lemma 3. Being given $\delta'\in]0,\ \delta'_0[$, δ satisfying (22), θ_0 and θ_1 in Θ such that $|\theta_0-\theta_1| \geq \delta'$, there exists a set $D_n(\theta_0,\theta_1,\delta')$ for which

$$(\forall\theta\in\Theta,\ |\theta-\theta_0| < \delta),\ P_\theta^{(n)}(D_n) \ge 1 - \exp\{-\tfrac{n}{4}\, Q[\bar{h}(\delta)]^2\}$$

$$(\forall\theta\in\Theta,\ |\theta-\theta_1| < \delta),\ P_\theta^{(n)}(D_n) \le \exp\{-\tfrac{n}{4}\, Q[\bar{h}(\delta)]^2\}.$$

XI.

There is no trouble considering, in the preceding inequalities, δ as a function of n, say $\delta = \delta_n$. The only precaution would be to distinguish the index n appearing in lemma 3 from the "current" index which is running between 1 and n in the proof of this lemma.

The sequence δ_n and the corresponding terms δ'_n, related by (22), being arbitrarily chosen (under the single condition $\delta'_n < \delta'_0$), we can cover Θ by open balls $B(\theta_i,\delta_n)$ $(i = 1,2,\ldots,K_n)$. Furthermore, it is always possible to suppose

$$K_n \le C\,\delta_n^{-1}$$

where C is a constant depending on Θ. For each $i = 1,2,\ldots,K_n$ we define:

$$I(i) = \{j\in\{1,2,\ldots,K_n\}/|\theta_j-\theta_i| \geq \delta'_n\} .$$

We put

$$\Delta_i = \bigcap_{j\in I(i)} D_n(\theta_i,\theta_j,\delta_n) \qquad \text{if} \quad I(i) \neq \phi$$

and

$$\Delta_i = X^{(n)} \qquad \text{if} \quad I(i) \neq \phi.$$

The set $I(i)$ contains at most (K_n-1) elements. So, as a direct consequence of lemma 3, we have, for all i:

$$(\forall\theta\in B(\theta_i,\delta_n)) \; P_\theta^{(n)}(\Delta_i) \geq 1 - (K_n-1)\exp\{-\tfrac{n}{4} Q[\overline{h}(\delta_n)]^2\} \tag{23}$$

and

$$(\forall\theta\in\Theta - B(\theta_i,\delta_n+\delta'_n)) \; P_\theta^{(n)}(\Delta_i) \leq \exp\{-\tfrac{n}{4} Q[\overline{h}(\delta_n)]^2\} . \tag{24}$$

Let θ_0 be an arbitrary fixed point in Θ. We define an estimator $\hat{\theta}_n(x^{(n)})$ as follows:

- if $x^{(n)} \notin \bigcup_1^{K_n} \Delta_i$, then $\hat{\theta}(x^{(n)}) = \theta_0$.

- if $x^{(n)} \in \bigcup_1^{K_n} \Delta_i$, put $i^* = \min\{i/1 \leq i \leq K_n, \; x^{(n)} \in \Delta_i\}$

and then

$$\hat{\theta}_n(x^{(n)}) = \theta_{i^*}.$$

It is evident that $\hat{\theta}_n$ is measurable, because it remains constant on Borelian sets of $X^{(n)}$.

XII.

Finally, we study the convergence of $\hat{\theta}_n$. For that, let θ be an arbitrary point in Θ. By the definition of $\hat{\theta}_n$, we can write:

$$(|\hat{\theta}_n(x^{(n)}) - \theta| > \delta_n+\delta'_n) \Rightarrow (x^{(n)} \notin \bigcup_1^{K_n} \Delta_i) \cup (|\theta_{i*}-\theta| > \delta_n+\delta'_n)$$

$$\Rightarrow (x^{(n)} \notin \bigcup_1^{K_n} \Delta_i) \cup (x^{(n)} \in \bigcup_{i:|\theta_i-\theta|>\delta_n+\delta'_n} \Delta_i).$$

Boole's inequality gives us, for every $\theta\in\Theta$:

$$P_\theta^{(n)}(|\hat{\theta}_n(X^{(n)})-\theta| > \delta_n+\delta'_n) \leq P_\theta^{(n)}(X^{(n)} \notin \bigcup^{K_n} \Delta_i)$$

$$+ \sum_{i:|\theta_i-\theta| > \delta_n+\delta'_n} P_\theta^{(n)}(\Delta_i). \qquad (25)$$

Inequality (24) implies

$$\sum_{i:|\theta_i-\theta| > \delta_n+\delta'_n} P_\theta^{(n)}(\Delta_i) \leq (K_n-1)\exp\{-\frac{n}{4} Q[\overline{h}(\delta_n)]^2\}. \qquad (26)$$

Besides, θ having any given value of Θ, there exists at least one point θ_r $(1 \leq r \leq K_n)$ such that $\theta\in B(\theta_r,\delta_n)$, and this legitimates the majoration:

$$P_\theta^{(n)}(X^{(n)} \notin \bigcup_1^{K_n} \Delta_i) \leq P_\theta^{(n)}(X^{(n)} \notin \Delta_r)$$

$$\leq (K_n-1)\exp\{-\frac{n}{4} Q[\overline{h}(\delta_n)]^2\}. \qquad (27)$$

Using (26) and (27), we deduce from (25):

$$(\forall\theta\in\Theta),\ P_\theta^{(n)}(|\hat{\theta}_n(X^{(n)})-\theta| > \delta_n+\delta_n') \leq 2(K_n-1)\exp\{-\tfrac{n}{4}\,Q[\overline{h}(\delta_n)]^2\}.$$

Remembering that $K_n \leq C\ \delta_n^{-1}$, we arrive at:

$$(\forall\theta\in\Theta),\ P_\theta^{(n)}(|\hat{\theta}(X^{(n)})-\theta| > \delta_n+\delta_n') \leq 2C\ \delta_n^{-1}\ \exp\{-\tfrac{n}{4}\,Q[\overline{h}(\delta_n)]^2\}.$$

It is rather obvious that we can choose δ_n converging to 0 at a sufficiently slow rate to insure:

$$\lim_{n\to\infty} \delta_n^{-1}\ \exp\{-\tfrac{n}{4}\,Q[\overline{h}(\delta_n)]^2\} = 0.$$

Then, (28) shows that $\hat{\theta}_n$ is satisfying the assumption (a) of theorem 3.

XIII.

The assumption (b) deserves may be a more explicit treatment. Putting $\psi(\delta_n) = \frac{1}{4}\,Q[\overline{h}(\delta_n)]^2$, it is sufficient to prove that we can choose $\delta_n \downarrow 0$ such that the series $u_n = \delta_n^{-1}\ \exp[-n\psi(\delta_n)]$ is convergent. Let α be a given positive number. By taking it small enough, it is possible to find δ_n' satisfying (22) if $\delta_n\in]0,\alpha[$.

Assuming this condition for α, let us consider, for each value of the integer m $(= 1,2,\ldots)$ the series:

$$u_{n,m} = (\tfrac{\alpha}{m})^{-1}\ \exp[-n\psi(\tfrac{\alpha}{m})].$$

It is convergent, because $\psi(\frac{\alpha}{m}) > 0$. So there exist integers $n_1, n_2, \ldots, n_m, \ldots$ such that

$$1 < n_1 < n_2 < \ldots < n_m < \ldots$$

and

$$(\forall m \geq 2) \quad \sum_{n=1+n_{m-1}}^{\infty} u_{n,m} < m^{-2}.$$

Let us define the sequence δ_n by:

$$\delta_n = \alpha \qquad \text{if} \quad 1 \leq n \leq n_1$$

$$\delta_n = \alpha/2 \qquad \text{if} \quad 1+n_1 \leq n \leq n_2$$

. . . .

$$\delta_n = \alpha/m \qquad \text{if} \quad 1+n_{m-1} \leq n \leq n_m,$$

and so on. We see easily that:

$$(\forall m \geq 2) \quad \sum_{n=1}^{n_m} u_n \leq \sum_{n=1}^{\infty} u_{n,1} + \sum_{n=1+n_1}^{\infty} u_{n,2} + \ldots + \sum_{n=1+n_{m-1}}^{\infty} u_{n,m}$$

$$\leq \sum_{n=1}^{\infty} u_{n,1} + 2^{-2} + 3^{-2} + \ldots + m^{-2}$$

$$< \sum_{n=1}^{\infty} u_{n,1} + \sum_{1}^{\infty} n^{-2}$$

from which the convergence of $\sum u_n$ is evident.

Remark. It is not difficult to extend theorem 3 to cases where Θ is unbounded, but with loss of the stochastic uniformity properties.

REFERENCES

[1] Geffroy, J. (1976). Inégalités pour le niveau de signification et la puissance de certains tests reposant sur des données quelconques. Comptes-Rendus Acad. Sc. (France), série A, t. 282, p. 1299-1301.

[2] Geffroy, J. (1976). Conditions suffisantes de convergence de certains tests déduits d'observations non nécessairement indépendantes ou équiréparties. In Recent developments in statistics, Barra J.R. et al., editors; North Holland Press, 1977.

[3] Geffroy, J. and Moché, R. (1974). Sur la séparation asymptotique uniforme des produits de lois de probabilité. Comptes-Rendus Acad. Sc. (France), série A, t. 278, p. 969-971.

[4] Geffroy, J. and Moché, R. (1975). Construction d'estimateurs uniformement convergents à partir d'observations non nécessairement équidistribuées. Comptes-Rendus Acad. Sc. (France), série A, t. 280, p. 133-135.

[5] Hillion, A. (1976). Sur l'integrale d'Hellinger et la séparation asymptotique. Comptes-Rendus Acad. Sc. (France), série A, t. 283, p. 61-64.

[6] Hillion, A. (1977). Quelques inégalités sur les régions de confiance. Ann. Inst. Henri-Poincaré (Paris), vol. XIII, n.4, p. 371-384.

[7] Hillion, A. (1978). Quelques questions de distances de lois de probabilité. Pub. Inst. Stat. Univ. de Paris, vol. XXIII, fasc. 3-4, p. 37-88.

[8] Hoeffding, W. and Wolfowitz, J. (1958). Distinguishability of sets of distributions. Ann. Math. Statist., vol. 29, p. 700-718.

[9] Moché, R. (1975). Construction d'estimateurs uniformément convergents. Comptes-Rendus Acad. Sc. (France), série A, t. 280, p. 1029-1031.

[10] Moché, R. (1977). Décantation et séparation asymptotique uniforme; tests et estimateurs convergents dans le cas d'observations indépendantes, équidistribuées ou non. Part of Doctorate Thesis, Univ. des Sciences et Techniques de Lille (France).

[11] Nemetz, T. (1974). Equivalence-orthogonality dichotomies of probability measures. Coll. on limit theorems of probability theory and statistics, Keszthely (Hungary).

[12] Rényi, A. (1967). On some problems of statistics from the point of view of information theory. Proc. of Coll. Information Theory, Debrecen, p. 343-357.

DENSITY ESTIMATION: ARE THEORETICAL RESULTS USEFUL IN PRACTICE?

B. W. Silverman

School of Mathematics
University of Bath
Bath, England

INTRODUCTION

Theoreticians in statistics often lose sight of the basic aim of statistics, the analysis of data. One of the most fundamental statistical concepts is the probability density function, and estimates of this function are of enormous importance in data analysis. Probability density estimates have been studied by an enormous number of authors, but there has been little attempt to reconcile theory and practice. In this paper, some reasons for estimating densities are discussed, after which the possible application of theoretical work to provide a practical method for determining how much to smooth a given data set is considered. The method ultimately obtained is objective and is applied to some engineering data.

DENSITY ESTIMATION AS A DATA ANALYTIC METHOD

Before going on to consider methods for probability density estimation in detail, it may be worth reviewing our reasons for estimating densities at all! Following the philosophy expounded by Tukey and others, but perhaps adding one additional stage, the process of analysing data can be divided into three sections:

ISBN: 0-12-166650-6

exploratory, confirmatory and presentational.

In exploratory data analysis, the technique of density estimation is of great importance. A detailed exposition of this and other aspects of density estimation is contained in Boneva, Kendall and Stefanov (1971). When confronted with a new data set, the statistician should use density estimates just as he uses other exploratory aids, to draw tentative conclusions about the data and to identify interesting features which may be investigated further later on. An example is the "mast cell" data considered by Emery and Carpenter (1974). The density estimates presented in Figure 1 suggested that the population represented by A was a mixture of the B population and a shifted population. This conclusion was reinforced by the consideration of non-parametric estimates of the density ratio (discriminant function); see Silverman (1978c) for details.

One point which may be worth stressing, though it perhaps goes against the spirit of the discussion below on the choice of smoothing parameter, is that density estimates of the same data smoothed by different amounts give emphasis to different features of the data. Thus, though there may be considerable value in being able to focus attention on one particular estimate as being correctly smoothed, it may be very valuable to note that other features become clear in other estimates. An excellent example is the Buffalo Snowfall data considered by Parzen (1979) and Tukey (1977), among others. This population appears unimodal at some smoothing scales and trimodal at others. While secondary modes which appear may not be statistically significant in any sense, the spirit of exploratory data analysis should lead us to examine further the observations giving rise to them to see if they are accounted for by some exogenous factor.

It should be pointed out that density estimation cannot be used as a 'back of an envelope' exploratory technique in the way that many of Tukey's techniques can. However, the advent of interactive computers has brought methods - such as density estimation - involving moderate amounts of computing easily within the reach of the exploratory data analyst.

The role of density estimation in confirmatory analysis is perhaps not as great. However, there are examples where the probability density estimate is needed in a confirmatory framework. Hermans and Habbema (1976) and others have used density estimates in the context of non-parametric discrimination. Another application of particular interest in view of recent work in a variety of areas (for example, Besag and Diggle (1977), Kendall (1974) and Kendall and Kendall, (1979)) is in simulating independent sets of data drawn from the same (unknown) distribution as a given collection of observations. A possible method is to estimate the unknown density and then to simulate from this density, perhaps using rejection sampling. If kernel density estimates are being used, one can perform the equivalent procedure of simulating random variables having the kernel as probability density function, scaling these by the window width, and adding the resulting quantities to randomly chosen members of the original data set. The procedure adopted by Kendall and Kendall (1979), of randomly perturbing each member of the given data set, can be thought of as a form of stratified sampling from the kernel density estimate.

An often neglected aspect of data analysis is presentation. It is easily forgotten how important it is for the statistician - whatever method he has used to analyse the data - to be able to present his results convincingly to the layman. Probability density functions are one of the only statistical devices that non-statisticians can comprehend, and so probability density estimates are of great importance in presentation. Indeed, it could be argued that this purpose alone justifies the study of probability density estimation.

METHODS FOR DENSITY ESTIMATION

Numerous methods have been suggested for estimating a probability density function from independent identically distributed observations. One of the first methods was the 'kernel method' introduced by Rosenblatt (1956); for real densities the estimate f_n is defined by

$$f_n(x) = \sum_{i=1}^{n} (nh)^{-1} \delta\{h^{-1}(x-X_i)\}$$

where $X_1,\ldots,X_n$ are i.i.d. observations, δ is a kernel function satisfying various conditions (e.g. $\delta \geq 0$, $\int \delta = 1$) and $h(n)$ is a sequence of 'window widths'. The kernel estimate has much to recommend it. It is intuitively appealing and easily understood. It is easy to compute in practice; one can either calculate it directly or perform the often more efficient procedure of discretising the data on a very fine grid and then performing the convolution using the Fast Fourier Transform. Finally, the simple explicit form of the estimator makes it much more mathematically tractable than most other methods, and so far more is known theoretically about its behaviour than for other estimators. Some of this theoretical work is considered below.

I would venture to suggest that some general principles hold when estimating densities. Firstly, most methods involve several choices of parameters, functions or functionals. In the kernel method one has to choose the kernel and the window width. However, except for the choice which controls the amount by which the data is smoothed, the resulting estimate is relatively robust to most of the choices; thus the choice of kernel in the kernel method has little effect of the efficiency of the method (see Epachenikov (1969)). However, the choice of smoothing parameter, in our case the window width, is of crucial importance because of the usual trade off between bias and random error; this choice has to be made, at least implicitly, in almost all methods. One of the most important problems in density estimation is the choice of smoothing parameter in whatever method is being used, a point eloquently made by Parzen (1979). This choice is probably more important even than the comparison between different methods.

SOME THEORETICAL RESULTS

There has been an enormous amount of theoretical work carried out on the kernel density estimate. For a review, see, for example, Rosenblatt (1971); recent important papers include those by Bickel and Rosenblatt (1973) and Bertrand-Retali (1978). Unfortunately, however, there has been little attempt to apply this theoretical work in practice; it is the contention of this paper that many of these theoretical results have genuine practical value.

One technique, used by Woodroofe (1967), Bickel and Rosenblatt (1973) and Silverman (1978a) is the use of results on the embedding of the empirical distribution function such as those of Brillinger (1969) and Komlos, Major and Tusnady (1974). These results can be applied to show that the random error in the kernel density estimate is approximately a Gaussian process. This fact follows easily from noticing that the estimate is a convolution of one kernel with the empirical distribution function, which is approximately the sum of the true distribution and a suitable version of the Brownian bridge. The convolution of the kernel with the Brownian bridge is the required Gaussian process.

The notation used by Silverman (1978a) gives considerable intuitive insight. It is possible to write down the following fundamental decomposition of the estimate:

$$f_{n,h} = f + b_h + n^{-1/2}\rho_h + \varepsilon_{n,h} \quad .$$

Here f is the true density, b_h is the bias, or systematic error, $n^{-1/2}\rho_h$ is a Gaussian process with zero mean and known variance/ covariance structure, and $\varepsilon_{n,h}$ is a secondary random error which is, under a wide range of conditions, asymptotically negligible compared with ρ. The elegance of the fundamental decomposition lies in the fact that b_h and ρ_h (in probability structure) depend only on the window width h and not on the sample size n.

Tearing apart the dependence on n and h in this way is of considerable value both intuitively and analytically because the behaviour of the estimate can be predicted from the properties of b and ρ, which both depend on only the single parameter h. Thus, for example, the effect of varying h for any fixed sample is easy to visualise conceptually. The process b and ρ respectively decrease and increase in magnitude as h decreases; under certain regularity conditions, $\sup|b|$ is exactly $\underline{O}(h^2)$ while $\sup|\rho|$ is exactly $\underline{O}[h^{-1/2}\{\log(1/h)\}^{1/2}]$ as $h \to 0$. For proofs see Bartlett (1963) and Silverman (1976). Thus the fact that bias and random error must be balanced to get optimal estimates becomes very clear.

The fundamental decomposition can be used to study the asymptotic consistency properties of the estimates. It is possible, for example, to show that, under mild conditions on the kernel δ, if f is uniformly continuous, and $h \to 0$ and $nh/\log n \to \infty$ as $n \to \infty$ that

$$\sup|f_n - f| \to 0 \qquad \text{a.s.} \quad \text{as} \quad n \to \infty.$$

Furthermore exact rates of convergence in this result can be obtained; for details see Silverman (1978a). A consequence of these rates of convergence is the following result; for exact details of proof and conditions see Silverman (1978b).

Theorem: Suppose f has uniformly continuous second derivative and δ satisfies various mild conditions. Suppose the sequence of window widths h is chosen to ensure the most rapid possible uniform convergence in probability of f_n to f.
Then, as $n \to \infty$

$$\sup(f_n'' - Ef_n'') \xrightarrow{p} k \sup|f''|$$

and

$$\inf(f_n'' - Ef_n'') \xrightarrow{p} -k \sup|f''|$$

where the constant k is given by

$$k = 1/2 \left|\int x^2\delta(x)dx\right| \{\int(\delta'')^2 dx/ \int \delta^2 dx\}^{1/2}.$$

This theorem can be used to obtain a method for choosing the window width appropriate to a given sample in practice, a problem discussed in the next section. This practical method will depend on considering various plots called 'test graphs' and so the theorem will be referred to subsequently as the test graph theorem.

CHOOSING THE WINDOW WIDTH IN PRACTICE

The significance of the test graph theorem can be appreciated by careful consideration of the various quantities involved. The term $f_n'' - Ef_n''$ represents the random error or noise in the second derivative f_n'' of the density estimate. Thus, if the window width is chosen optimally for the estimation of f, the noise in f''_n will be $\underline{O}(1)$ as n tends to infinity. Furthermore, the asymptotic value of $\sup|f''_n - Ef''_n|$ is the product of a known constant k with $\sup|f''|$; since Ef''_n, the trend in f''_n, is asymptotically equal to f'', it follows that the conclusion of the test graph theorem can be reexpressed by saying that for optimal estimation of f, the noise in f''_n should have magnitude approximately k times the magnitude of the trend of f''_n.

The kernel we shall use here is the piecewise quartic

$$\delta(x) = \begin{cases} \frac{1}{4}|x|^4 - \frac{1}{2}|x|^3 + \frac{1}{2} & \text{for } |x| \le 1 \\ \frac{1}{4}|x|(2-|x|)^3 & \text{for } 1 \le |x| \le 2 \\ 0 & \text{otherwise} \end{cases}$$

For this kernel the constant k is almost exactly 0.4. (For the Gaussian kernel the constant is slightly smaller, 0.306.) Thus, the test graph theorem predicts that for optimal estimates, the noise in f''_n will be of magnitude comparable with but somewhat smaller than the trend of f''_n.

The suggested practical method based on the test graph theorem is to draw "test graphs" of the second derivative of the density estimate for various window widths. Choose the window width which gives fluctuations of the right size in the test graph, and then use this window width to construct the estimate of the true density. In the actual calculation, the test graph can either be expressed as

$$f''_n(x) = n^{-1}h^{-3} \sum_{i=1}^{n} \delta''\{h^{-1}(x-X_i)\}$$

or else, if Fourier transform methods are being used, the Fourier transform of f_n can be divided by $-t^2$ to give the Fourier transform of the test graph f''_n.

For the moment, we shall consider the assessment of the test graphs subjectively, though an objective method of choice will be discussed below. The method using subjective choice has previously been described more fully in Silverman (1978b). The subjective choice of test graph is carried out by applying the remarks made above to obtain the principle that

> "the ideal test graph should have fluctuations which are quite marked but do not obscure the systematic variation completely."

Although subjective application of this principle would appear at first sight to present difficulties, in practice the character of the test graph changes rapidly as the window width is varied - indeed far more rapidly than the corresponding density estimate - and so the choice of an appropriate window width is easily made.

Before going on to discuss some applications, it should be stressed that the method, and in particular the choice of the second derivative of the density estimate for consideration, is not an ad hoc procedure but has a firm, if asymptotic, theoretical justification.

SOME ILLUSTRATIONS AND APPLICATIONS

In order to investigate the test graph method, it is first applied to simulated data. Since the underlying density is known, it is of course possible to find empirically the best window width, in the sense of minimising absolute uniform error. The first example illustrates the principle, and is constructed from 100 observations simulated from a standard Normal distribution. Test graphs for various window widths are shown in Figure 2. The fluctuations in the graph for window width 0.5 clearly overwhelm the systematic variation, while the test graph for window width 0.9 is a smooth curve where systematic variation clearly dominates. The best window width is 0.675 in this case, and the corresponding test graph shows what is meant by "marked fluctuations which do not obscure the systematic variation."

Further insight is given by considering a much larger sample of 10,000 observations again from a Normal density. For the particular sample under consideration, the optimal window width was found to be 0.27; three of the test graphs are shown in Figure 3. The change in character between these graphs is quite striking and demonstrates clearly the behavior predicted by the test graph principle. Thus the highly theoretical test graph theorem presented above leads to properties which are of real value when estimating the density in practice.

Next we consider the application of the principle to a data set arising in practice. This example appeared previously in Silverman (1978b), and is some circular data of M.A. Stephens, consisting of 76 observations of the orientation of turtles. This data set has previously been analysed by Mardia (1975) and Boneva, Kendall and Stefanov (1971) among others. It is easily shown that the test graph theorem holds on the circle. Some of the test graphs are shown in Figure 4. Those for window widths 20° and 45° are clearly unacceptable; comparison with the test graphs for Normal data and application of the test graph principle led to the choice of 32° as the ideal window width. The density estimate for this window width is shown in Figure 5. The bimodal nature of the data is clearly demonstrated, though the second mode appears to be somewhat weaker than some previous analyses have suggested. Once the data have been presented in this way, confirmatory analysis of the data scarcely seems necessary, since all relevant conclusions (bimodality, modes 180° apart) are now obvious.

Another example, concerning some simulated data, is considered in Silverman (1978b) and will not be reproduced here. A third example will be discussed below, in the discussion of methods for assessing the test graphs automatically.

One of the advantages of the kernel method of density estimation over other methods is its easy and natural generalization to the multivariate case. While it is in many circumstances perhaps inappropriate to estimate densities in high dimensional space, estimates in the bivariate and possibly trivariate cases are clearly of considerable practical value. The test graph can be generalized to the multivariate case; for details the reader is referred to Silverman (1978b). Another generalization, due to Gratton (1979), is the estimation not of the density itself, but of some function (such as the logarithm) of the density. Again a test graph procedure can be applied but the definition of the test graph must be modified.

CHOOSING THE WINDOW WIDTH OBJECTIVELY

Although some experience - by others as well as by the author - has shown that the subjective assessment of the test graph usually provides a procedure which is quite satisfactory, some would say that it is preferable to have an objective method for assessing the test graphs and hence choosing the appropriate window width. Again our preference will be for a method with some theoretical justification, and so we shall first discuss the problem theoretically. Decompose the second derivative of the density estimate as

$$f_n'' = s + r$$

where s is the systematic term Ef_n'' and r is the random part $f_n'' - Ef_n''$.

Recall that the test graph principle required that r should be slightly smaller in magnitude than s. Consider, now, the total variation $V(f_n'')$ of the test graph. As the window width h is varied, $V(f_n'')$ will vary; if h is large compared with its optimal value, s will dominate r, and so the behavior of $V(f_n'')$ will be determined by the behavior of $V(s)$. If h is too small, on the other hand, the behavior of $V(f_n'')$ will be determined by the behavior of $V(r)$ since s will be small compared with r.

It will be seen that the behavior of $V(s)$ and $V(r)$ are quite different and so it will usually be possible to detect the change in the behavior of $V(f_n'')$ and hence find the point where r and s are comparable. Only brief statements of results will be given here; for details see Silverman (1979).

It can be shown theoretically that under certain regularity conditions $V(r)$ is asymptotically proportional to $n^{-1/2}\, h^{-7/2}$, so that for fixed (large) sample size

$$\frac{d \log V(r)}{d \log h} \doteq -\frac{7}{2} .$$

On the other hand it can be shown that

$$\frac{d \log V(s)}{d \log h} \to 0 \quad \text{as} \quad h \to 0$$

and by consideration of tractable special cases that for plausible values of h this quantity seems unlikely to go outside the range $(-1.5, 0)$. Hence the problem of selecting the appropriate test graph can be reduced to that of finding a change in the slope of $\log V(f_n'')$ plotted against $\log h$ from -3.5 to a larger value, probably between

0 and -1.5. The easiest way of doing this is to find the minimum of, say, $\log V(f_n'') + 2.5 \log h$, since the change will not generally take place abruptly.

Since the derivation of the objective method of assessing the test graphs is partly heuristic and partly asymptotic, it is important to check by simulation that the method works in practice. A small scale simulation study has indicated that, in the majority of cases, the value of h which minimises the criterion function $\log V(f_n'') + 2.5 \log h$ is very close to the optimal window width, though it would perhaps be wise to examine the test graphs themselves as a final check.

The best illustration of the method is perhaps by reference to some data obtained by Adrian Bowyer in the study of surface roughness of stainless steel. The data may be considered to consist of fifteen thousand observations of the height of a steel surface above an arbitrary datum, and were thought of as independent identically distributed random variables. The calculations were performed by fast Fourier transform methods and the Gaussian kernel was used. The resulting criterion function is shown in Figure 6. The minimising value of h is in fact 0.55, and the corresponding test graph is shown in Figure 7. The author's subjective preference is for a slightly smoother test graph, such as that for window width 0.625 shown in Figure 8, but it is almost impossible to detect the difference in the corresponding density estimates. The estimate for window width 0.55 is shown in Figure 9.

The density estimate shows that the height distribution has a long lower tail but a short upper tail, and that the main part of the distribution is correspondingly skewed. This accords with the engineer's conception of the surface as having originally had a symmetric height distribution, the upper tail having been modified by abrasion which

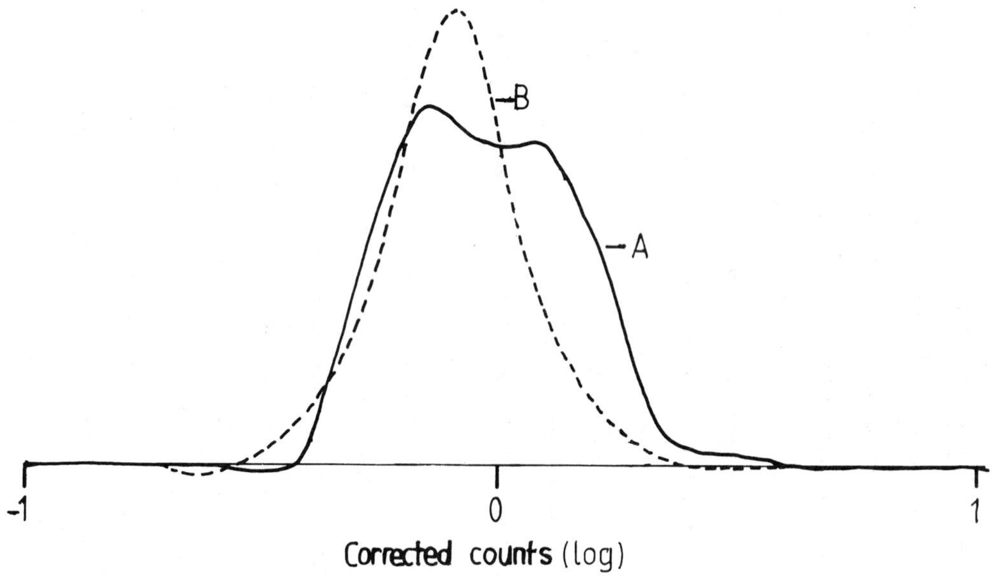

Figure 1. Corrected mast cells counts. A: Sudden deaths; B: Hospital deaths.

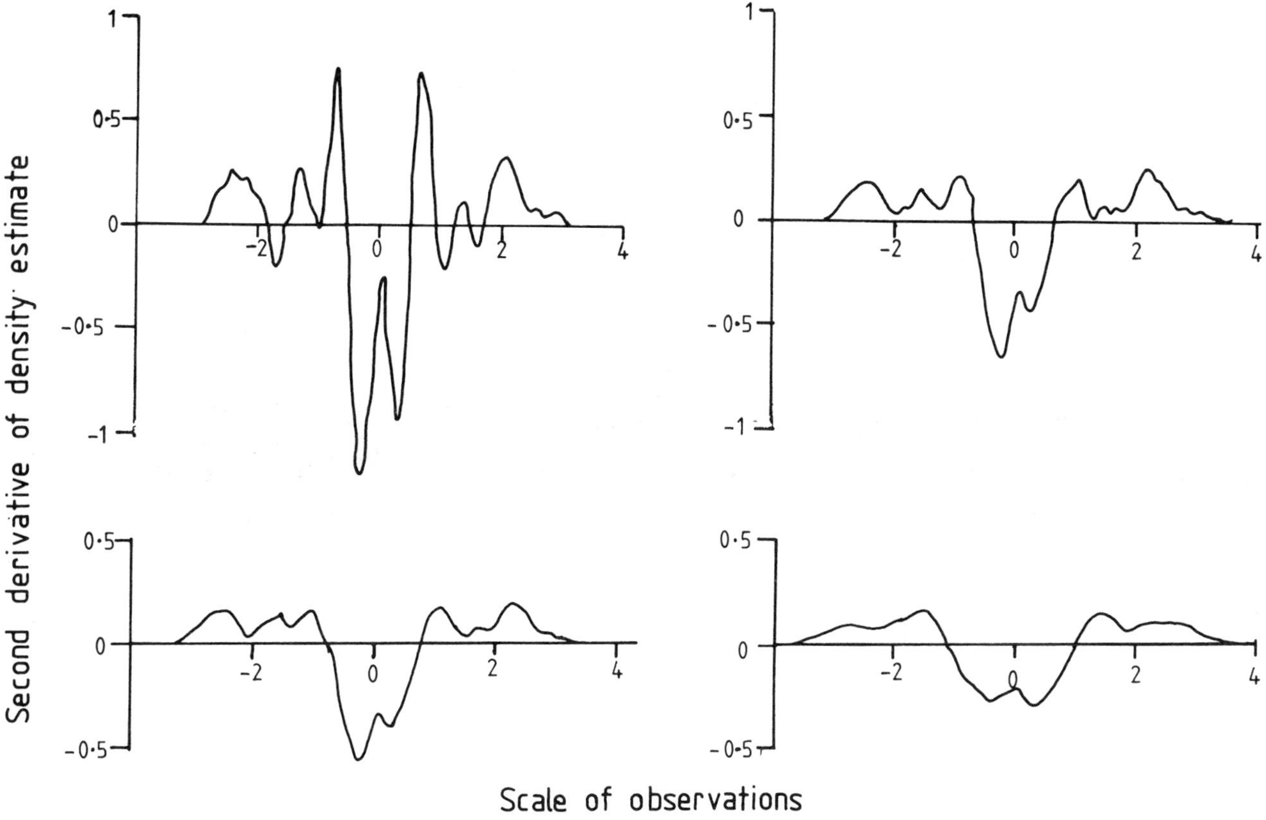

Figure 2. Test graphs for 100 Normal observations. Window widths; top left, 0.5; top right, 0.625; bottom left, 0.675; bottom right, 0.9.

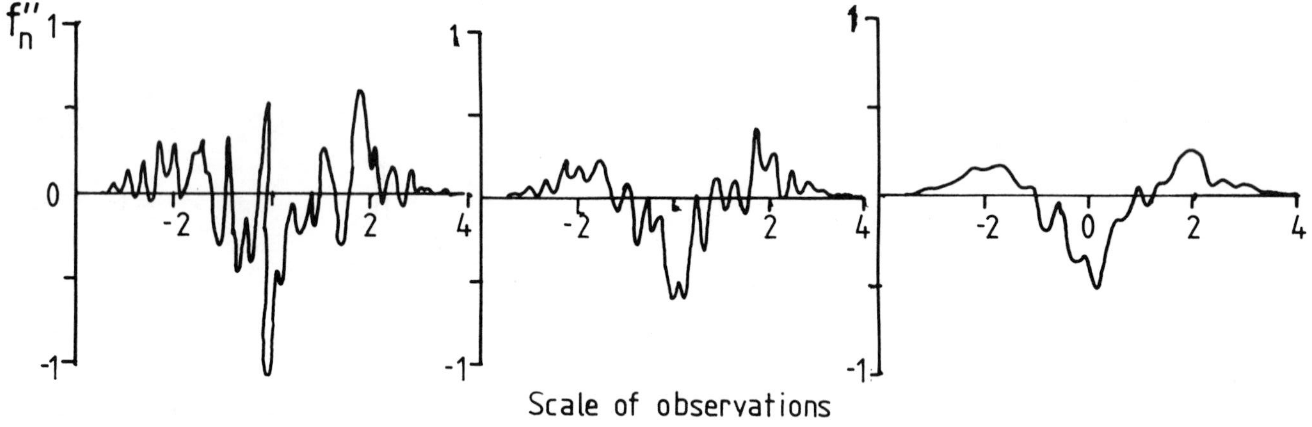

Figure 3. Test Graphs for 10,000 Normal Observations. Window widths, left to right, 0.21, 0.27, 0.35.

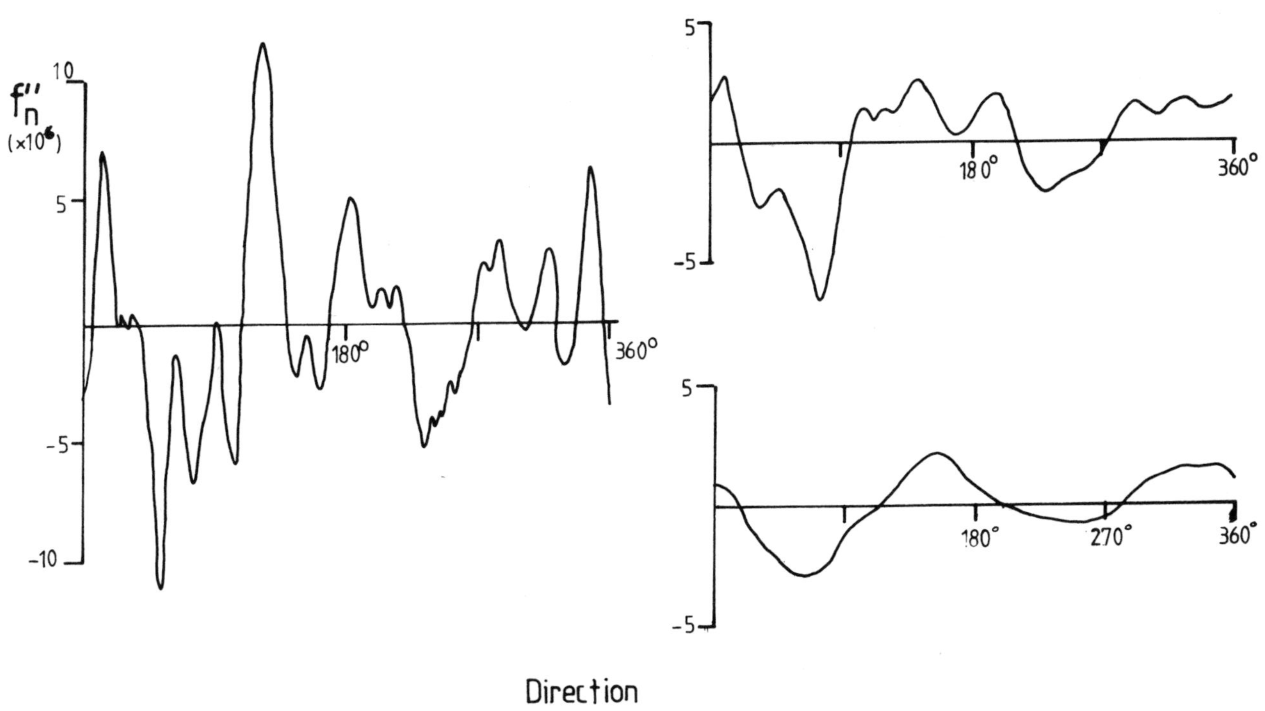

Figure 4. Test graphs for turtle data. Window widths: left, 20°; top right, 32°; bottom right, 45°.

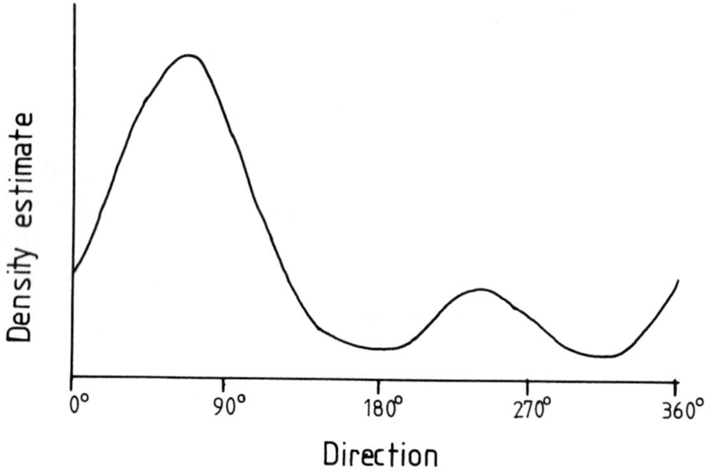

Figure 5. Density estimate for turtle data, window width 32°.

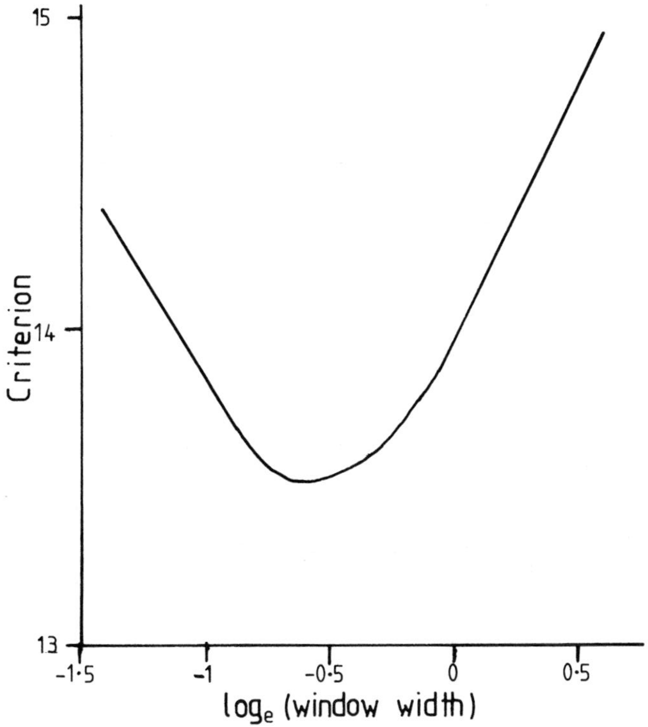

Figure 6. Criterion function for steel data.

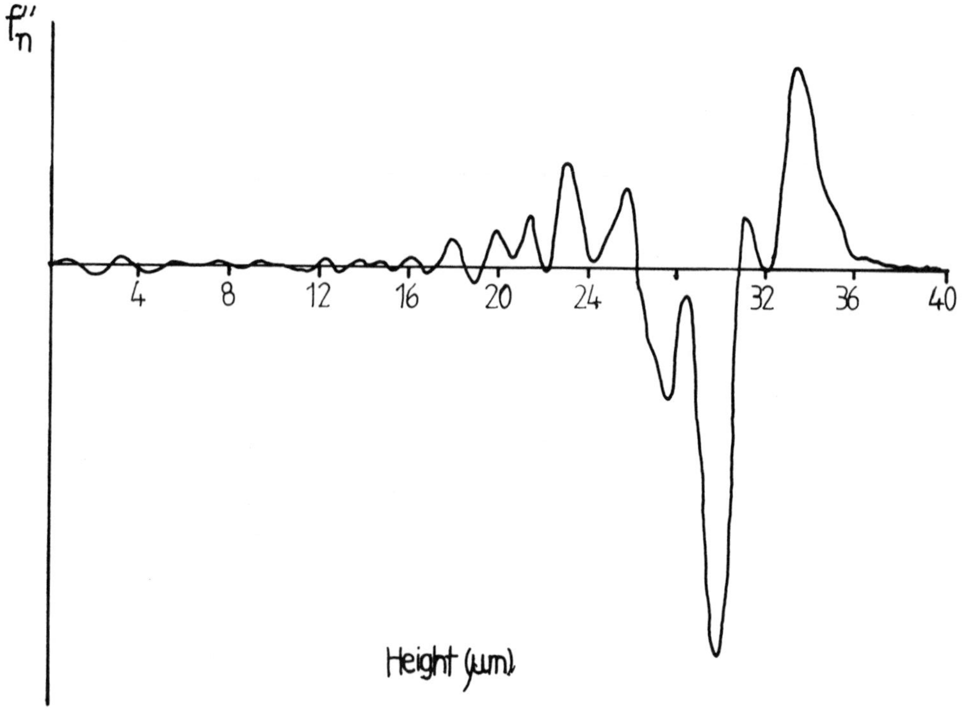

Figure 7. Test graph for steel data, window width 0.55.

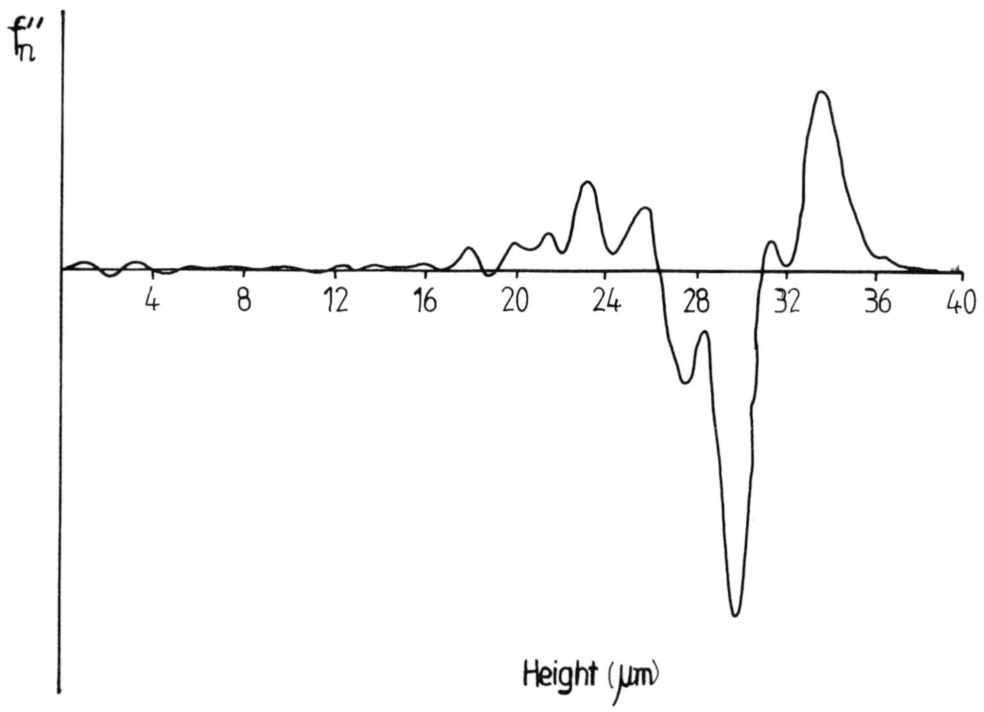

Figure 8. Test graph for steel data, window width 0.625.

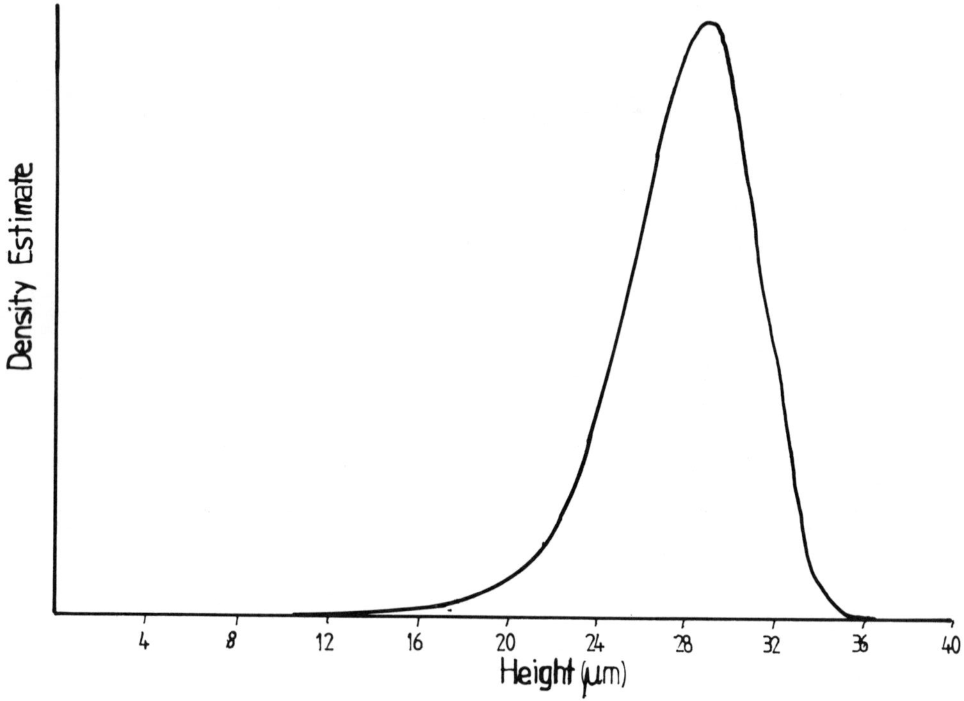

Figure 9. Density estimate for steel data, window width 0.55.

tends to remove the 'hills' on the surface while leaving the 'valleys' unchanged. It should also be noted that the lower tail is somewhat heavier than that of a Normal distribution.

CONCLUSION AND ACKNOWLEDGMENTS

Although some of the theoretical work that has been carried out for density estimation seems to have little possible practical value, it appears that asymptotic results can be used to provide a solution to a most important practical problem. Perhaps there is a lesson here for theoretical <u>and</u> practical statisticians; some effort to understand each other may be richly repaid!

I am most grateful to several colleagues for helpful discussion and comments, and to Adrian Bowyer for his help with some of the computing.

REFERENCES

Bartlett, M. (1963). Statistical estimation of density functions, Sankhya A, 25, 245-54.

Bertrand-Retali, M. (1978). Convergence uniforme d'un estimateur de la densité par la méthode de noyau, Rev. Roum. Math. Pures et Appl, 23, 361-85.

Besag, J.E. & Diggle, P.J. (1977). Monte Carlo tests for spatial pattern, Appl. Statist., 26, 327-33.

Bickel, P.J. & Rosenblatt, M. (1973). On some global measures of the deviations of density function estimates, Ann. Statist., 1, 1071-95.

Boneva, L.I., Kendall, D.G. & Stefanov, I. (1971). Spline transformations, J. Roy. Statist. Soc. B, 33, 1-70.

Brillinger, D. (1969). An asymptotic representation of the sample distribution function, Bull. Amer. Math. Soc., 75, 545-7.

Emery, J.L. and Carpenter, R.G. (1974). Pulmonary mast cells in infants and their relation to unexpected death in infancy. Proceedings of the Francis E. Camps International Symposium on Sudden and Unexpected Deaths in Infancy, Canadian Foundation for the Study of Infant Deaths.

Epachenikov, V.A. (1969). Nonparametric estimation of a multivariate probability density, Theory Prob. Applic., 14, 153-8.

Gratton, R.J. (1979). Generalizing the test graph method for estimating functions of an unknown density, Biometriak, 66,

Habbema, J.D.F., Hermans, J. and van den Broek, K. (1974). A stepwise discriminant analysis program. COMPSTAT 1974, Physica Verlag, Wien.

Kendall, D.G. (1974). Hunting quanta, Phil. Trans. Roy. Soc. Lond. A, 276, 231-66.

Kendall, D.G. & Kendall, W.S. (1979). Internal alignments in random two-dimensional sets of points.

Komlos, J., Major, P. & Tusnady, G. (1975). An approximation of partial sums of independent random variables, Zeit. fur Wahr., 32, 111-31.

Mardia, K.V. (1975). Statistics of directional data, J. Roy. Statist. Soc. B, 37, 349-93.

Parzen, E. (1979). Nonparametric statistical data modelling, J. Amer. Stat. Ass., 74,

Rosenblatt, M. (1956). Remarks on some nonparametric estimates of a density function, Ann. Math. Statist., 27, 832-7.

Rosenblatt, M. (1971). Curve estimates, Ann. Math. Statist., 42, 1815-42.

Silverman, B.W. (1976). On a Gaussian process related to multivariate probability density estimation, Math. Proc. Camb. Phil. Soc., 80, 134-144.

Silverman, B.W. (1978a). Weak and strong uniform consistency of the kernel estimate of a density and its derivatives, Ann. Statist., 6, 177-84.

Silverman, B.W. (1978b). Choosing a window width when estimating a density, Biometrika, 65, 1-11.

Silverman, B.W. (1978c). Density ratios, empirical likelihood and cot death, Appl. Statist., 27, 26-33.

Silverman, B.W. (1979). On estimating densities automatically. (in preparation)

Tukey, J.W. (1977). Exploratory Data Analysis, Addison-Wesley, New York.

Woodroofe, M. (1967). On the maximum deviation of the sample density, Ann. Math. Statist., 38, 475-81.

STABILITY THEOREMS FOR CHARACTERIZATIONS OF THE NORMAL AND OF THE DEGENERATE DISTRIBUTIONS

Eugene Lukacs*

The Catholic University
Washington, D.C.

I. INTRODUCTION

A great variety of assumptions can be used to characterize probability distributions (populations). One of those most frequently used is a regression property which we formulate next.

Let X and Y be a two random variables and suppose that the expectation $E(Y)$ and Y exists. The random variable Y is said to have constant regression with respect to X if the relation

$$E(Y|X) = E(Y) \tag{1.1}$$

holds almost everywhere. If $E(Y) = 0$ then we say that Y has zero regression on X.

The following characterization theorem is known.

<u>Theorem A</u>. Let X_1, X_2 be independently and identically distributed random variables and suppose that $E(X_1) = 0$. Let $L_1 = a_1X_1 + a_2X_2$ and $L_2 = b_1X_1 + b_2X_2$ be two linear forms in these random variables and assume that L_1 has zero regression on L_2.

---

*The preparation of this paper was supported by the National Science Foundation under grants MCS 577-01834 and MCS 577-01834 A01.

ISBN: 0-12-166650-6

(a) If $|a_2/a_1| < 1$ or if $|a_2/a_1| = 1$ but $|b_2/b_1| < 1$ then the random variables X_1 and X_2 have degenerate distributions.

(b) Assume further that $E(X_1^2) < \infty$ and that $a_1b_1 + a_2b_2 = 0$ and $|b_2/b_1| < 1$ then X_1 and X_2 are normally distributed.

Note 1° It is obviously no restriction to assume that $|b_2/b_1| \leq 1$; in the following we will always make this assumption

2° If $E(X^2) \neq 0$ and if (b) is satisfied then the distribution of the X_j is a non-degenerate normal distribution.

Theorem A is due to C.R. Rao ([3] theorem 1).

In this paper we wish to derive two stability theorems which correspond to theorem A. That is, we shall suppose that the assumptions of theorem A are only approximately satisfied and we will investigate how this modification affects the conclusions of the thorem. We shall see that the conclusions are then also only approximately satisfied. For such a study one must precisely define the meaning of the word "approximately"; this is our next task.

Definition 1. Let X and Y be two random variables and suppose that the expectation $E(Y)$ of Y exists. We say that Y has almost zero regression on X, or more concisely, that Y has ϵ-regression on X if $|E(Y|X)| \leq \epsilon$ almost everywhere.

To define almost normality of a distribution function we introduce a metric in the space of distribution functions. This can be done in a variety of ways; for the discussion of our problem it is convenient to use a metric introduced by P. Lévy.

Definition 2. Let F and G be two distribution functions. Their distance $L(F,G)$ is defined by the formula

$$L(F,G) = \inf\{h: F(x-h) - h \leq G(x) \leq F(x+h) + h\}. \tag{1.2}$$

It can be shown (see for instance B.V. Gnedenko - A.N. Kolmogorov [1]) that $L(F,G)$ defines a metric in the space of distribution functions. Convergence in this metric is equivalent to the weak convergence of distribution functions. This fact indicates the importance of the Lévy metric for the study of probabilistic problems.

In this paper we denote the degenerate distribution by $E(x)$, that is

$$E(x) = \begin{cases} 0 & \text{if } x < 0 \\ 1 & \text{if } x \geq 0 \end{cases}$$

and write

$$\Phi(x) = \frac{1}{\sqrt{2\pi}} \int_{-\infty}^{x} e^{-y^2/2}\, dy$$

for the standardized normal distribution.

We can now formulate our results.

Theorem 1. Let X_1 and X_2 be two independently and identically distributed random variables with common distribution function $F(x)$ and suppose that the first moment of F exists and that $\mathcal{E}(X_1) = 0$. Let $L_1 = a_1X_1 + a_2X_2$ and $L_2 = b_1X_1 + b_2X_2$ be two linear forms in the random variables X_1 and X_2 and suppose that L_1 has ϵ-regression on L_2. We write $\alpha = a_2/a_1$, $\gamma = b_2/b_1$ and assume that either $|\alpha| < 1$ or $|\alpha| = 1$ but $|\gamma| < 1$. Then

$$L(F,E) \leq C\, \epsilon^{(1-2\delta)/3} \tag{1.3}$$

where $0 < \delta < \frac{1}{2}$ and where C is given by formulae (3.5a), (3.13) and (3.18).

Theorem 2. Let X_1 and X_2 be two independently and identically distributed random variables with common distribution function $F(x)$ and suppose that the second moment of $F(x)$ exists and that $E(X_1) = 0$ while $0 < E(X_1^2) < \infty$. Let again $L_1 = a_1X_1 + a_2X_2$ and $L_2 = b_1X_1 + b_2X_2$ be two linear forms and suppose that $a_1b_1 + a_2b_2 = 0$ while $|\gamma| = |b_2/b_1| < 1$. We assume that L_1 has ϵ-regression on L_2. Then there exists a normal distribution $\Phi_2 = \Phi(\frac{x-c_2}{\sigma_2})$ such that

$$L(F,\Phi_2) = K(\ell n \tfrac{1}{\epsilon})^{-1/11} \tag{1.4}$$

Remark: It is no restriction to assume that $|\gamma| \leq 1$.

In both theorems we assume that L_1 has ϵ-regression on L_2, that is

$$|E(L_1|L_2)| \leq \epsilon.$$

This can be written in the form

$$E(L_1|L_2) = R \tag{1.5}$$

where R is a random variable such that

$$|R| \leq \epsilon \text{ almost everywhere.} \tag{1.5a}$$

According to the conditions of the theorems $E(X_j) = 0$, $j = 1,2$, so that we see from (1.5) that

$$E(R) = E(L_1) = 0. \tag{1.5b}$$

We introduce the function

$$k(t) = E(\mathrm{Re}\, e^{itL_2}) \tag{1.6}$$

and see easily that

$$k(0) = 0 \tag{1.7a}$$

$$k(-t) = \overline{k(t)} \tag{1.7b}$$

$$|k(t)| \leq \epsilon. \tag{1.7c}$$

In section 2 we bring some lemmas needed for the proofs of both theorems, in section 3 theorem 1 is proven. Section 4 contains three lemmas used in proving theorem 2 while the proof of theorem 2 is given in section 5.

II. LEMMAS NEEDED FOR THE PROOF OF BOTH THEOREMS

Inequality (1.7c) follows immediately from (1.6), and gives an estimate for the function $k(t)$. However a better estimate is available for small values of t.

Lemma 2.1. The estimate

$$|k(t)| \leq \beta_1(|b_1| + |b_2|)\ |t|\ \epsilon$$

holds. Here $\beta_1 = \int_{-\infty}^{\infty} |x|dF(x)$ is the absolute moment of order 1 of the X_j.

We use the MacLaurin expansion of the exponential function $e^{itL_2} = 1 + \theta itL_2$ (with $|\theta| \leq 1$) and substitute it into (1.6) and get

$$k(t) = E[R(1 + \theta itL_2)]$$

or, in view of (1.5b) $k(t) = E(it\theta RL_2)$ so that

$|k(t)| \leq \epsilon |t| \; E(|L_2|) \leq \epsilon |t| \; (|b_1| + |b_2|) \; \beta_1$. Lemma 2.1 is proven.

Lemma 2.2. Let $f(t)$ be, a possibly complex valued, continuous function of the real variable t and suppose that $|f(0)| \neq 0$ and $|f(-t)| = |f(t)|$. Let $0 < \epsilon < |f(0)|$ and suppose that

$$q(\epsilon) = \sup\{c: |f(t)| > \epsilon \text{ for } |t| < c\} < \infty.$$

Assume further that $|f(t)| > \eta$ for $|t| < q(\epsilon)$ then $\epsilon > \eta$. For the proof of lemma 2.2 we refer to Lukacs [2].

The next lemma estimates the Lévy-distance between two distribution functions in terms of the difference between their characteristic functions.

Lemma 2.3. Let $F(x)$ and $G(x)$ be two distribution functions and denote their characteristic functions by $f(t)$ and $g(t)$ respectively. Let

$$\Delta = \sup_{t>0} |f(t) - g(t)|t^{-\lambda} \qquad (\lambda > 0)$$

then we can estimate the Lévy distance $L(F,G)$ by

$$L(F,G) \leq A_\lambda \; \Delta^{1/(1+\lambda)}$$

where the constant $A_\lambda < [2(1 + \lambda)]^2 \; [\lambda\pi^{1/(1+\lambda)}]^{-1}$.

Lemma 2.3 is a particular case of a result of V.M. Zolotarev [6], [7].

III. PROOF OF THEOREM 1.

It follows from formulas (1.5) and (1.6) that

$$E(L_1 e^{itL_2}) = k(t),$$

or written in greater detail

$$a_1 E(X_1 e^{itb_1X_1})E(e^{itb_2X_2}) + a_2 E(e^{it_1b_1X_1})E(X_2 e^{itb_2X_2}) = k(t). \quad (3.1)$$

Let $f(t) = \int_{-\infty}^{\infty} e^{itx} dF(x)$ be the characteristic function of the distribution function $F(x)$ (of the random variables X_j, $j = 1,2$). The expectations in formula (3.1) can be expressed in terms of $f(t)$ and $f'(t)$ thus one obtains the equation

$$a_1 f'(b_1t)f(b_2t) + a_2 f(b_1t)f'(b_2t) = i\, k(t). \quad (3.2)$$

There exists a t-interval I in which $f(b_1t) \neq 0$ and $f(b_2t) \neq 0$; for values $t \in I$ we can define $\phi(t) = \log f(t)$ and we obtain from (3.2) the relation

$$a_1 \phi'(b_1t) + a_2 \phi'(b_2t) = \frac{i\, k(t)}{f(b_1t)f(b_2t)}.$$

We write $v = b_1t$ and obtain from the last equation

$$\phi'(v) + \alpha\, \phi'(\gamma v) = A(v) \quad (3.3)$$

where

$$A(v) = \frac{i}{a_1} \frac{k(\frac{v}{b_1})}{f(v)\, f(\gamma v)} \text{ and } \gamma = \frac{b_2}{b_1},\ \alpha = \frac{a_2}{a_1},\ |\gamma| \leq 1. \quad (3.3a)$$

It follows from (3.3) that

$$\alpha\, \phi'(\gamma v) + \alpha^2\, \phi'(\gamma^2 v) = \alpha\, A(\gamma v).$$

We repeat this procedure and add the equations so obtained and see that

$$\phi'(v) + \alpha^{n+1}\, \phi'(\gamma^{n+1} v) = \sum_{j=0}^{n} \alpha^j\, A(\gamma^j v). \tag{3.4}$$

Let

$$q = q(\epsilon) = \sup\{\xi : |f(v)| > \epsilon^\delta \text{ for } |v| < \xi\} \tag{3.5}$$

where the constant δ is selected so that

$$0 < \delta < \frac{1}{2}.$$

Since $|\gamma| \le 1$ one has $|f(v)| > \epsilon^\delta$ and $|f(\gamma v)| > \epsilon^\delta$ for $|v| < q$. From now on we restrict the values of v to the interval $|v| < q$.

We consider here only the case where $q < \infty$, the proof is carried in the same way if $q = \infty$ and is therefore not repeated in this paper.

We use formula (3.3a) and lemma 2.1 to get an estimate for $A(t)$.

$$|A(t)| \le \frac{|b_1|+|b_2|}{|a_1|}\, \beta_1 |t|\, \epsilon^{1-2\delta} \qquad (|t| < q)$$

or if we write

$$C_1 = (|b_1| + |b_2|)\, \beta_1 / |a_1| \tag{3.5a}$$

$$|A(t)| \le C_1 |t|\, \epsilon^{1-2\delta} \qquad (|t| < q). \tag{3.5b}$$

Therefore

$$|\alpha^j A(\gamma^j v)| \leq C_1 |\alpha\gamma|^j \; |v| \; \epsilon^{1-2\delta} \qquad (|v| < q). \tag{3.6}$$

It follows from the assumptions of theorem 1 that either $|\alpha| < 1$ and $|\gamma| \leq 1$ or $|\alpha| = 1$ and $|\gamma| < 1$, moreover, $\phi'(0) = 0$. Therefore one has

$$\lim_{n\to\infty} \alpha^{n+1} \phi'(\gamma^{n+1} v) = 0. \tag{3.7}$$

Moreover $|\alpha\gamma| < 1$ and one concludes from (3.6) that the series

$$\sum_{j=0}^{\infty} \alpha^j A(\gamma^j v)$$

converges absolutely and uniformly if $|v| < q$. In view of (3.7) we see from (3.4) that

$$\phi'(v) = \sum_{j=0}^{\infty} \alpha^j A(\gamma^j v) \qquad (|v| < q). \tag{3.8}$$

Therefore

$$\phi(t) = \sum_{j=0}^{\infty} \alpha^j \int_0^t A(\gamma^j v)dv \qquad (|v| < q)$$

or

$$f(t) = \exp\left\{\sum_{j=0}^{\infty} \alpha^j \int_0^t A(\gamma^j v)dv\right\} \qquad (|t| < q). \tag{3.9}$$

We estimate the sum in the exponent of (3.9):

$$\left|\sum_{j=0}^{\infty} \alpha^j \int_0^t A(\gamma^j v)dv\right| \leq \sum_{j=0}^{\infty} |\alpha|^j \left|\int_0^t A(\gamma^j v)dv\right|. \tag{3.10}$$

We have, if $t > 0$ and $|t| < q$

$$\left|\int_0^t A(\gamma^j v)dv\right| \leq \int_0^t C_1|\gamma|^j\, |v|\, \epsilon^{1-2\delta}dv = C_1|\gamma|^j\, \epsilon^{1-2\delta}\int_0^t v\, dv \quad (|t| < q)$$

so that

$$\left|\int_0^t A(\gamma^j v)dv\right| \leq C_1|\gamma|^j\, \epsilon^{1-2\delta}\, t^2/2 \qquad (|t| < q). \tag{3.11}$$

By a similar reasoning one can show that (3.11) is also valid if $t < 0$. It follows from (3.10) and (3.11) that

$$\left|\sum_{j=0}^{\infty} \alpha^j \int_0^t A(\gamma^j v)dv\right| \leq C_1\, \epsilon^{1-2\delta}\, \frac{t^2}{2} \sum_{j=0}^{\infty} |\alpha\gamma|^j \qquad (|t| < q).$$

Since $|\alpha\gamma| < 1$, we see that

$$\left|\sum_{j=0}^{\infty} \alpha^j \int_0^t A(\gamma^j v)dv\right| \leqq \frac{C_1}{2[1-|\alpha\gamma|]}\, \epsilon^{1-2\delta}\, t^2 \qquad (|t| < q). \tag{3.12}$$

We note that

$$e^{-|w|} \leq |e^w| \leq e^{|w|}\ .$$

We write

$$C_2 = C_1[1 - |\alpha\gamma|]^{-1}/2 \tag{3.13}$$

and combine (3.9) and (3.12) and get for $|t| < q$

$$\exp\left\{-C_2\, \epsilon^{1-2\delta}\, t^2\right\} \leq |f(t)| \leq \exp\left\{C_2\, \epsilon^{1-2\delta}\, t^2\right\}\ .$$

Hence

$$|f(t)| > \exp\left\{-C_2\ \epsilon^{1-2\delta}\ q^2\right\} \quad \text{for} \quad |t| < q$$

and also, by the definition of q,

$$|f(t)| > \epsilon^{\delta} \quad \text{for} \quad |t| < q.$$

It follows therefore from lemma 2.2 that

$$\epsilon^{\delta} > \exp[-C_2\ \epsilon^{1-2\delta}\ q^2]$$

hence

$$q^2 > \frac{\delta}{C_2}\ \frac{\log \frac{1}{\epsilon}}{\epsilon^{1-2\delta}}\ . \tag{3.14}$$

We see from equation (3.9) that

$$|f(t) - 1| = \left|\exp\left\{\sum_{j=0}^{\infty} \alpha^j \int_0^t A(\gamma^j v)dv\right\} - 1\right| \qquad (|t| < q)$$

or, since $|e^w - 1| \leq |w|\ e^{|w|}$,

$$|f(t)-1| \leq \left|\sum_{j=0}^{\infty} \alpha^j \int_0^t A(\gamma^j v)dv\right| \exp\left\{\left|\sum_{j=0}^{\infty} \alpha^j \int_0^t A(\gamma^j v)dv\right|\right\} .$$

Using (3.12) and (3.13) we see that

$$\left|\frac{f(t)-1}{t^2}\right| \leq C_2\ \epsilon^{1-2\delta} \exp(C_2\ \epsilon^{1-2\delta}\ t^2) \qquad (|t| < q). \tag{3.15}$$

Let ξ, $\frac{1}{2} < \xi < 1$ be a constant to be selected later and define $Q > 0$ by

$$Q^2 = \frac{\xi}{C_2\ \epsilon^{1-2\delta}} < \frac{\delta}{C_2} \log \frac{1}{\epsilon} /\epsilon^{1-2\delta} < q^2. \qquad (3.16)$$

(The first inequality in (3.16) holds if ϵ is so small that $\delta \log \frac{1}{\epsilon} > 1$).

From now on we restrict t to the interval $|t| < Q$. Using (3.15) and (3.16) we have

$$\left|\frac{f(t)-1}{t^2}\right| \leq \frac{\xi}{Q^2}\ e^{\xi} \quad \text{for} \quad |t| < Q. \qquad (3.17)$$

We apply now lemma 2.3 with $\lambda = 2$. Then

$$L(F,E) \leq A_2\ \Delta^{1/3}$$

where $A_2 < 9\pi^{-1/3}$ while

$$\Delta = \sup_{t>0} \left|\frac{f(t)-1}{t^2}\right|.$$

Let

$$\Delta_1 = \sup_{0 < t < Q} \left|\frac{f(t)-1}{t^2}\right| \quad \text{and} \quad \Delta_2 = \sup_{t \geq Q} \left|\frac{f(t)-1}{t^2}\right|$$

then $\Delta = \max(\Delta_1, \Delta_2)$.

Since $\Delta_2 \leq \frac{2}{Q^2}$ we see from (3.17) that

$$\Delta \leq \max\left(\frac{\xi e^{\xi}}{Q^2}, \frac{2}{Q^2}\right).$$

The equation $xe^x = 2$ has a single positive root in the interval $(\frac{1}{2}, 1)$. Let x_0 be this root and put $\xi = x_0$. Then

$$\Delta \leq 2/Q^2 = 2C_2\, \epsilon^{1-2\delta}/x_0$$

and

$$L(F,E) \leq 9\pi^{-1/3}\, (2C_2/x_0)^{1/3}\, \epsilon^{(1-2\delta)/3} = C\, \epsilon^{(1-2\delta)/3}$$

where

$$C = 9\left(\frac{2C_2}{x_0\pi}\right)^{1/3}. \tag{3.18}$$

Theorem 1 is proven.

IV. LEMMAS USED IN THE PROOF OF THEOREM 2

We see from (1.5) that $L_2R = E(L_1L_2|L_2)$, therefore

$$E(L_2R) = E(L_1L_2). \tag{4.1}$$

In theorem 2 we assumed that the second moment of $F(x)$ exists and that its first moment is zero. The existence of the second moment implies the existence of $E(L_2R)$ so that the function $k(t)$, defined by (1.6), as $k(t) = E(Re^{itL_2})$ can be differentiated. Lemma 4.1 asserts the existence of this derivative and gives an estimate for it.

Lemma 4.1. If the second moment $\sigma^2 = \int_{-\infty}^{\infty} x^2\, dF(x)$ of the random variables X_j exists and if we suppose that $\int_{-\infty}^{\infty} x\, dF(x) = 0$ and that the coefficients of the linear forms L_1 and L_2 satisfy the relation $a_1b_1 + a_2b_2 = 0$ then the function $k(t)$, defined by (1.6) can be differentiated and $|k'(t)| \leq (b_1^2 + b_2^2)\, \sigma^2 |t| \epsilon$.

The differentiability of $k(t)$ has already been established so that we have only to prove the estimate of the lemma.

We have

$$k'(t) = i\, E(RL_2\, e^{itL_2}) \tag{4.2}$$

and we use again the MacLaurin expansion of $e^{itL_2} = 1 + \nu itL_2$, $(|\nu| \leq 1)$ and see that

$$k'(t) = i\, E(RL_2) - \nu t\, E(RL_2^2).$$

In view of the assumption that $E(X_j) = 0$ $(j = 1,2)$ one has from (4.1)

$$E(RL_2) = E(L_1L_2) = (a_1b_1 + a_2b_2)\sigma^2 = 0 \tag{4.3}$$

so that

$$|k'(t)| \leq |t|\, E(RL_2^2) \leq |t|\epsilon\, E(L_2^2) = (b_1^2 + b_2^2)\sigma^2|t|\epsilon \tag{4.4}$$

and the lemma is proven.

We also see from (4.2) and (4.3) that

$$k'(0) = 0. \tag{4.5}$$

<u>Lemma 4.2</u>. Let $F(x)$ and $G(x)$ be two distribution functions and write $F_a(x) = F(\frac{x}{a})$, $G_a(x) = G(\frac{x}{a})$ where $a > 0$. Then $\min(a,1)\ L(F,G) \leq L(F_a,G_a) \leq \max(a,1)\ L(F,G)$. The lemma is due to J.W. Thompson [4].

<u>Lemma 4.3</u>. Let $\Phi(x)$ be the standardized normal distribution and let $F(x)$ be a decomposable distribution with factors F_1 and F_2 i.e. $F = F_1 * F_2$ such that $L(F,\Phi) \leq \epsilon$. Then there exists a normal distribution $\Phi_1(x) = \Phi(\frac{x-a_1}{\sigma_1})$ such that $L(F_1,\Phi_1) \leq C[\ell n \frac{1}{\epsilon}]^{-1/11}$. Here C is a numerical constant, independent of ϵ. The lemma is due to V.M. Zolotarev [5].

V. PROOF OF THEOREM 2

We had

$$\phi'(v) + \alpha^{n+1}\ \phi'(\gamma^{n+1}\ v) = \sum_{j=0}^{n} \alpha^j\ A(\gamma^j v). \tag{3.4}$$

In view of the assumptions of theorem 2 this relation can be differentiated and one obtains

$$\phi''(v) + (\alpha\gamma)^{n+1}\ \phi''(\gamma^{n+1}\ v) = \sum_{j=0}^{n} (\alpha\gamma)^j\ A'(\gamma^j v).$$

Since $a_1 b_1 + a_2 b_2 = 0$ we see that $\alpha\gamma = -1$ and get

$$\phi''(v) + (-1)^{n+1}\ \phi''(\gamma^{n+1} v) = \sum_{j=0}^{n} (-1)^j\ A'(\gamma^j v). \tag{5.1}$$

Our next task is the estimation of $A'(v)$. We introduce as in (3.5)

$$q = q(\epsilon) = \sup\{\xi:\ |f(t)| > \epsilon^{\delta} \quad \text{for} \quad |t| < \xi\} \tag{5.1a}$$

but select δ so that

$$0 < \delta < \frac{1}{3}$$

and restrict our considerations to the interval $|t| < q$. We see from formula (3.3a) that for $|t| < q$

$$\frac{a_1}{i} A'(v) = \frac{1}{b_1} \frac{k'(\frac{v}{b_1})}{f(v)f(\gamma v)} - \frac{k(\frac{v}{b_1})}{f(v)f(\gamma v)} \left[\frac{f'(v)}{f(v)} + \gamma \frac{f'(\gamma v)}{f(\gamma v)}\right]. \tag{5.2}$$

Using lemma 4.1 we get an estimate for the first term on the right hand side of (5.2):

$$\left|\frac{1}{b_1} \frac{k'(\frac{v}{b_1})}{f(v)f(\gamma v)}\right| \le \frac{(b_1^2+b_2^2)\sigma^2}{b_1^2} \epsilon^{1-2\delta}|v|. \tag{5.2a}$$

To estimate the second term on the right hand side of (5.2) we note that $\left|\frac{f'(a)}{f(a)}\right| \le E(|X|)\ \epsilon^{-\delta} = \beta_1\ \epsilon^{-\delta}$. Therefore

$$\left|\frac{f'(v)}{f(v)} + \gamma \frac{f'(\gamma v)}{f(\gamma v)}\right| \le \beta_1(1 + |\gamma|)\ \epsilon^{-\delta}\ .$$

Using lemma 2.1 we see that

$$\left|\frac{k(\frac{v}{b_1})}{f(v)f(\gamma v)} \left[\frac{f'(v)}{f(v)} + \gamma \frac{f'(\gamma v)}{f(\gamma v)}\right]\right| \le \frac{|b_1|+|b_2|}{|b_1|} \beta_1^2(1+|\gamma|)\epsilon^{1-3\delta}|v|. \tag{5.2b}$$

We combine (5.2), (5.2a) and (5.2b) and obtain

$$|A'(v)| \leq \frac{(b_1^2 + b_2^2)\sigma^2}{|a_1|b_1^2} \epsilon^{1-2\delta} |v| + \frac{|b_1|+|b_2|}{|a_1 b_1|} \beta_1^2(1 + |\gamma|)\epsilon^{1-3\delta}|v|.$$

We write

$$C_1 = (1 + \gamma^2)\sigma^2/|a_1|, \quad C_2 = [(1 + |\gamma|)\beta_1]^2/|a_1|$$

and

$$C = C_1 + C_2$$

then we can rewrite the estimate for $|A'(v)|$ as

$$|A'(v)| \leq C|v| \epsilon^{1-3\delta} \qquad (|v| < q) \tag{5.3}$$

so that

$$|A'(\gamma^j v)| \leq C|\gamma|^j |v| \epsilon^{1-3\delta} \qquad (|v| < q).$$

We had

$$k(-t) = \overline{k(t)} \tag{1.7b}$$

and see easily from (3.3a) that

$$A'(-v) = \overline{A'(v)}; \tag{5.4a}$$

moreover

$$\phi''(-t) = \overline{\phi''(t)}. \tag{5.4b}$$

We symmetrize (5.1) by writing

$$\begin{cases} \Psi(t) = \phi(t) + \phi(-t) \\ \Psi''(t) = \phi''(t) + \phi''(-t) \\ B(t) = A'(t) + A'(-t) \end{cases} \tag{5.5}$$

and obtain the equation

$$\Psi''(v) + (-1)^{n+1} \Psi''(\gamma^{n+1} v) = \sum_{j=0}^{n} (-1)^j B(\gamma^j v). \tag{5.6}$$

From (1.7a), (4.5) and (5.2) we see that

$$A'(0) = 0$$

hence also

$$B'(0) = 0.$$

We put $v = 0$ in (5.6) and obtain

$$\Psi''(0) + (-1)^{n+1} \Psi''(0) = 0. \tag{5.6a}$$

We subtract (5.6a) from (5.6) and get

$$\Psi''(v)-\Psi''(0)+(-1)^{n+1} [\Psi''(\gamma^{n+1} v)-\Psi''(0)] = \sum_{j=0}^{n} (-1)^j B(\gamma^j v). \tag{5.7}$$

The functions Ψ'' and B are real and we see from (5.3) that

$$|B(v)| \leq 2C|v| \; \epsilon^{1-3\delta} \quad \text{for} \quad |v| < q.$$

Therefore

$$|(-1)^j \; B(\gamma^j v)| \leq 2C|v| \; \epsilon^{1-3\delta} \; |\gamma|^j \quad (|v| > q).$$

Since $|\gamma| < 1$, we see that the infinite series $\sum_{j=0}^{\infty} (-1)^j B(\gamma^j v)$ is absolutely and uniformly convergent and is majorated by a series with constant terms.

We also note that

$$\lim_{n\to\infty} (-1)^{n+1}[\Psi''(\gamma^{n+1}v) - \Psi''(0)] = 0$$

and that

$$\Psi''(0) = -2\sigma^2.$$

We let $n \to \infty$ in (5.7) and obtain

$$\Psi''(v) = \sum_{j=0}^{\infty} (-1)^j \; B(\gamma^j v) - 2\sigma^2. \tag{5.8}$$

Suppose that $q > t \geq \tau$, then

$$\int_0^\tau [\sum_{j=0}^{\infty} (-1)^j \; B(\gamma^j v)]dv = \sum_{j=0}^{\infty} (-1)^j \int_0^\tau B(\gamma^j v)dv \tag{5.9}$$

where

$$|(-1)^j \int_0^\tau B(\gamma^j v)dv| \leq C\, \epsilon^{1-3\delta}|\gamma^j|\tau^2. \tag{5.9a}$$

The series on the right hand side of (5.9) can be integrated term by term and one obtains

$$S = \int_0^t \left\{ \int_0^\tau [\sum_{j=0}^{\infty} (-1)^j\; B(\gamma^j v)]dv \right\} d\tau = \sum_{j=0}^{\infty} (-1)^j \int_0^t [\int_0^\tau B(\gamma^j v)dv]d\tau. \tag{5.10}$$

By a similar reasoning it can be shown that (5.10) is also valid if $t < \tau < 0$ but $|t| < q$ so that (5.10) holds for all $|t| < q$.

It follows then that

$$\Psi(t) = -\sigma^2 t^2 + S. \tag{5.11}$$

We use now (5.9a) to get an estimate for S and see that

$$|S| \leq C\, \epsilon^{1-3\delta} \frac{|t|^3}{3(1-|\gamma|)} \qquad (|t| < q). \tag{5.12}$$

Let

$$h(t) = |f(t)|^2 = e^{\Psi(t)}$$

then

$$h(t) = \exp(-\sigma^2 t^2 + S) \qquad (|t| \leq q). \tag{5.13}$$

Since the relations $|f(v)| > \epsilon^{\delta}$ and $|h(v)| > \epsilon^{2\delta}$ are equivalent we can rewrite (5.1a) in the form

$$q = \sup\{\xi: \ |h(t)| > \epsilon^{2\delta} \ \text{for} \ |t| < \xi\}.$$

From (5.12) and (5.13) it follows that

$$|h(t)| \leq |e^S| \leq \exp(C_1 \epsilon^{1-3\delta} |t|^3) \qquad (|t| < q).$$

where $C_1 = C/[3(1 - |\gamma|)]$.

Therefore

$$|h(t)| \geq \exp(-C_1 \epsilon^{1-3\delta} |t|^3) \qquad (|t| < q).$$

We apply again lemma 2.2 and see that

$$\epsilon^{2\delta} > \exp(-C_1 \epsilon^{1-3\delta}|t|^3) \qquad (|t| < q)$$

hence,

$$q^3 > \frac{2\delta}{C_1} \frac{\log \frac{1}{\epsilon}}{\epsilon^{1-3\delta}} . \tag{5.14}$$

One also has from (5.13)

$$|h(t) - e^{-\sigma^2 t^2}| \leq |e^S| -1| \leq |S| \ e^{|S|} \qquad (|t| < q)$$

or

$$|h(t) - e^{-\sigma^2 t^2}| \leq C_1 \epsilon^{1-3\delta}|t|^3 \ \exp(C_1 \epsilon^{1-3\delta}|t|^3) \qquad (|t| < q).$$

so that

$$\left|\frac{h(t)-e^{-\sigma^2 t^2}}{t^3}\right| \leq C_1 \epsilon^{1-3\delta} \exp(C_1 \epsilon^{1-3\delta} |t|^3) \quad (|t| < q). \tag{5.15}$$

Again let x_0 be the unique positive root of the equation $xe^x = 2$ which is located in the interval $(\frac{1}{2}, 1)$ and put

$$Q^3 = \frac{x_0}{C_1 \epsilon^{1-3\delta}} . \tag{5.16}$$

For ϵ sufficiently small (i.e. so small that $\frac{2\delta}{x_0} \log \frac{1}{\epsilon} > 1$) one has

$$Q^3 < \frac{2\delta}{x_0} \log \frac{1}{\epsilon} Q^3 = \frac{2\delta}{C_1} \frac{\log \frac{1}{\epsilon}}{\epsilon^{1-3\delta}} < q^3$$

and, according to (5.15), for $|t| < Q$

$$\left|\frac{h(t)-e^{-\sigma^2 t^2}}{t^3}\right| \leq C_1 \epsilon^{1-3\delta} \exp(C_1 \epsilon^{1-3\delta} Q^3) = \frac{x_0}{Q^3} \epsilon^{x_0} .$$

It follows from the definition of x_0 that

$$\left|\frac{h(t)-e^{-\sigma^2 t^2}}{t^3}\right| \leq \frac{2}{Q^3} \quad \text{for} \quad |t| < Q.$$

We apply lemma 2.3 for $\lambda = 3$. We have $A_3 < 32(3\pi^{1/4})^{-1}$ and write $G(x) = \Phi(\frac{x}{\sigma\sqrt{2}})$ for the normal distribution with zero mean and variance $2\sigma^2$. Let $H(x)$ be the distribution function which corresponds to the characteristic function $h(t)$. It is then easily seen that

$$L(H,G) < 32(3\pi^{1/4})^{-1} \left(\frac{2}{Q^3}\right)^{1/4}$$

or if we write

$$C_2 = \frac{32}{3} \left(\frac{2C_1}{\pi x_0}\right)^{1/4} \tag{5.17a}$$

$$\epsilon_1 = \epsilon^{(1-3\delta)/4} \tag{5.17b}$$

we get

$$L(H,G) < C_2 \, \epsilon_1 = O(\epsilon^{(1-3\delta)/4}). \tag{5.17}$$

Since $h(t) = |f(t)|^2$ one concludes that

$$H = F * \tilde{F}$$

where $\tilde{F}(x) = 1 - F(x-0)$ is the distribution conjugate to $F(x)$. Let $G_1(x) = \Phi(\frac{x}{\sigma})$, then $G(x) = G_1 * G_1 = \Phi(\frac{x}{\sigma\sqrt{2}})$. We see from lemma 4.2 that

$$L(H(\tfrac{x}{a}), G(\tfrac{x}{a})) \le \max(a,1) \, L(H,G)$$

or putting $a = \frac{1}{\sigma\sqrt{2}}$ and noting that $G(x) = \Phi(\frac{x}{\sigma\sqrt{2}})$

$$L(H(x\,\sigma\sqrt{2}), \Phi(x)) \le C_3 \, \epsilon_1 \tag{5.18}$$

where

$$C_3 = C_2 \max\left(\frac{1}{\sigma\sqrt{2}}, 1\right).$$

The distribution function $H(x\,\sigma\sqrt{2})$ has the factor $F_1 = F(x\,\sigma\sqrt{2})$ and we wish to show that the distribution $F(x\,\sigma\sqrt{2})$ is close to a suitably chosen Normal distribution.

We see from (5.18) and lemma 4.3 that there exists a normal distribution $\Phi_1 = \Phi(\frac{x-c_1}{\sigma_1})$ such that

$$L(F_1,\Phi_1) \le C[\log(\frac{1}{\epsilon_1})]^{-1/11}.$$

According to (5.16b) $\epsilon_1 = \epsilon^{(1-3\delta)/4}$, hence $\log(\frac{1}{\epsilon_1}) = \frac{1-3\delta}{4}\log\frac{1}{\epsilon}$. We write

$$C_4 = C\,(\frac{1-3\delta}{4})^{-1/11}$$

and see that

$$L(F_1,\Phi_1) \le C_4(\log\frac{1}{\epsilon})^{-1/11}. \qquad (5.19)$$

We use again lemma 4.2 to show that there exists a normal distribution $\Phi_2 = \Phi(\frac{x-a_2}{\sigma_2})$ such that

$$L(F,\Phi_2) \le K(\log\frac{1}{\epsilon})^{-1/11} = O[(\log\frac{1}{\epsilon})^{-1/11}].$$

The choice of X_{c_1} and σ_1 (of a_2 and σ_2) is indicated in Zolotarev [5].

REFERENCES

[1] Gnedenko, B.V. and Kolmogorov, A.N. Limit distributions for sums of independent random variables. Addison-Wesley, Cambridge, Mass. 1954.

[2] Lukacs, E. Stability theorems for charactierizations by constant regression. Periodica Math. Hungarica 2(1972), p. 111-128.

[3] Rao, C.R. On some characterizations of the normal law. Sankhya Ser. A, 29(1967), p. 1-14.

[4] Thompson, J.W. A note on the Lévy distance. Journal Applied Probability 12(1975), p. 412-414.

[5] Zolotarev, V.M. On the stability of the decomposition of the normal law into components. Teoriya verojatn. i ee primen. 13(1968), p. 738-742 [English Translation: Theory of probability and applic. 13(1968), p. 697-700].

[6] Zolotarev, V.M. Some new inequalities in probability connected with the Lévy metric. Dokl. Akad. Nauk SSSR 190(1970), p. 1019-1021 [English Translation: Soviet Math. Dokl. 11, p. 231-234].

[7] Zolotarev, V.M. Estimates of the difference between distributions in the Lévy metric. Trudi Steklov 112(1971), 224-231 [English Translation: Proceedings of the Steklov Institute 112, p. 232-240].

ESTIMATION OF THE SUPPORT CONTOUR-LINE OF A PROBABILITY LAW: LIMIT LAW

Jacques Chevalier

Institut de Statistique
Université Pierre et Marie Curie
Paris, France

I. INTRODUCTION

In a former paper [2] we estimated from a sample the support contour-line Γ and the support K of a probability law P.

We came to study in this paper the following points

a) For support estimation

- Sufficient and necessary conditions of convergence in probability, almost sure (w.p.1) and almost complete (a.co.).
- Speed of convergence.

b) For support contour-line estimation

- Sufficient and necessary conditions of convergence (in probability, w.p.1, a.co.) when using the C^1 distance between curves.

We intend now to study limit laws on support estimation.

II. NOTATIONS AND HYPOTHESIS

A. d being the euclidian distance of R^2, we call $B(x,\epsilon)$ the open sphere of center x and radius ϵ. For all $\epsilon > 0$ and all sub-set A of R^2, we call ϵ-retraction and ϵ-dilatation of A the following sets

ISBN: 0-12-166650-6

$$A_\epsilon = \{M \mid d(M,\bar{A}) \geq \epsilon\}; \quad A^\epsilon = \{M \mid d(M,A) \leq \epsilon\} \tag{1}$$

with the convention $d(M,\phi) = +\infty$.

We shall call A ϵ-convex if

$$A = C(\bigcup_{x \in B} B(x,\epsilon)), \quad \text{with} \quad B = CA^\epsilon \tag{2}$$

B. P is a probability measure defined on R^2, with compact support K, absolutely continuous with respect to Lebesgue measure λ^2, and which density verifies the following condition:

$$\forall M \in K, \quad 0 < \alpha \leq \phi(M) \leq \beta \tag{3}$$

with real constants α and β. α and β may (or not) be known.

C. The frontier of K is a simple Jordan arc (closed, with no double point) Γ which is rectifiable. Let us call F the length of Γ, M_o a point from Γ and s the curvilinear representation of $[0,F]$ on Γ with origin M_o such that when t goes from 0 to F, $M = s(t)$ describes Γ in the positive direction. We assume that s admits a continuous second derivative and verifies

$$s'(0) = s'(F) \quad \text{and} \quad s''(0) = s''(F) \tag{4}$$

We know that for all $t \in [0,F]$, $s'(t)$ differs from the nul vector.

Let us call R_t the curvature radius (which can be infinite) of Γ at point $s(t)$. We know likewise that the existence of a real $S > 0$ such that

$$\forall\ t \in [0,F], \quad |R_t| \geq S \tag{5}$$

We orientate the perpendicular $\vec{N}_t$ to Γ towards the inside of Γ, such that the two vectors $(\vec{T}_t, \vec{N}_t)$ constitute a direct basis, with $\vec{T}_t = ds/dt$.

Let us call $\Gamma(u)$ the curve with representation

$$s_u(t) = s(t) + u\vec{N}_t \tag{6}$$

We know the existence of $u_o > 0$ such that for all $|u| \leq u_o$

- K is $|u|$-convex
- $\Gamma(u)$ is a simple Jordan arc
- s_u is of class C^1 and verifies $s_u'(0) = s_u'(F)$ and

$$\forall\ t \in [0,F] \quad s_u'(t) = 0 \tag{7}$$

- The application v_u: $s(t) \to s_u(t)$ is a bijection from Γ to $\Gamma(u)$
- If $u > 0$, $\Gamma(u)$ is the frontier of the retraction K_u from K, if $u < 0$, $\Gamma(u)$ is the frontier of the dilatation K^{-u} from K.

Set $\lambda(u)$ the difference between K and K_u if $u > 0$ (between K^{-u} and K if $u < 0$).

To obtain these results, we just have to notice that for any $t_o \in [0,F]$, $s(t_o)$ admits a neighbourhood V on Γ such that for $|u| < |R_{t_0}|$, the application v_u: $s(t) \to s_u(t)$ restricted to V is a bijection. (7) is realised when u_o is less than $S/2$. We end the demonstration of the above properties by using the compactness of Γ.

D. Consider a sequence of independent v.a. $(X_i)_{i \in N^*}$ from the same law P and a sequence $(\rho_n)_{n \in N^*}$ of strictly positive reals verifying

$$\rho_n = o(1); \qquad o(\rho_n^2) = Ln/n \tag{8}$$

Define

$$I_n = \{x | \forall \ i=1,\ldots,n, \quad X_i \notin B(x,\rho_n)\} \tag{9}$$

and

$$H_n = C \bigcup_{x \in I_n} B(x,\rho_n) \tag{10}$$

Call Γ_n the frontier of H_n and F_n the length of Γ_n. Q being a point from R^2, there exists at least a point Q' from Γ such that $d(Q,Q') = d(Q,\Gamma)$; QQ' is the perpendicular at Q' to Γ. Let us call v the application "projection" which associates to Q the set of points Q' verifying $d(Q,Q') = d(Q,\Gamma)$. The restriction of v to $\wedge(u_o) \cup \wedge(-u_o)$ is a univocal bijective application from Γ_n to Γ.

When Γ_n belongs to $\wedge(u_o) \cup \wedge(-u_0)$ (which according to [2], holds almost completely for n tending towards $+\infty$) the C^0 distance between Γ_n and Γ is equal to

$$\Delta(\Gamma_n,\Gamma) = \sup_{Q \in \Gamma_n} d(Q,\Gamma)$$

We want to study the limit law of $\Delta(\Gamma_n,\Gamma)$.

E. We shall use for the demonstration a theorem on rare events due to Geffroy [3].

<u>Theorem</u>. (Geffroy). Let E_{nr} be a double sequence of random events defined for all positive n and for $1 \le r \le k(n)$ where $k(n)$ is a given function with integer values growing indefinitely with n. Assume that the sequence E_{nr} verifies

(A) For all value of n and all set $(r_1, r_2, \ldots, r_\nu)$ such that $1 \le r_1 < r_2 < \ldots < r_\nu \le k(n)$ we have

$$P\left(\bigcap_{i=1}^{\nu} E_{nr_i}\right) \le \prod_{i=1}^{\nu} P(E_{nr_i})$$

(B) In the same hypothesis, the real $\epsilon_{nr_1 \ldots r_\nu}$ defined by

$$P\left(\bigcap_{i=1}^{\nu} E_{nr_i}\right) = \left(\prod_{i=1}^{\nu} P(E_{nr_i})\right)(1 + \epsilon_{nr_1 \ldots r_\nu})$$

verifies the relation $\epsilon_{nr_1 \ldots r_\nu} \le \epsilon(n)$ where $\epsilon(n)$ is a function of n with limit 0 with $1/n$.

(C) $\max\limits_{1 \le r \le k} P(E_{nr}) \underset{n}{\to} 0$

In these conditions, if the sequence $\sigma_n = \sum_{r=1}^{k} P(E_{nr})$ admits a limit θ when $n \to +\infty$

$$\lim_n P\left(\bigcup_{r=1}^{k} E_{nr}\right) = 1 - \exp(-\theta)$$

III. 1st CASE: ϕ IS CONSTANT

Suppose ϕ constant and equal to c on K. Let $k(n)$ be a sequence of integers satisfying

$$k(n) \to +\infty;\ k(Lk)^2 = o(n);\quad n = o(k^3\rho_n) \tag{11}$$

Taking z real, set:

$$z_n = \frac{k}{n} z + \frac{k\ Lk}{ncF} \tag{12}$$

and

$$z'_n = \frac{k}{n} z + \frac{k\ Lk}{ncF_n} \tag{13}$$

Then we have the following result

Theorem 1. For any $z \in R$,

$$\underline{\lim}_n\, P(\Delta(\Gamma_n,\Gamma) < z_n) \geq \exp(-\exp(-cFz)) \tag{14}$$

and if,

$$k\ Lk(Ln/n\rho_n^2)^{2/3} = o(1) \tag{15}$$

$$\underline{\lim}_n\, (P(\Delta(\Gamma_n,\Gamma) < z'_n)/\exp(-\exp(-czF_n))) \underset{\text{a.co.}}{\geq} 1 \tag{16}$$

Remark: (11) and (15) can be satisfied both; for instance we can take

$$\rho_n = n^{-1/6};\quad k(n) = n^{2/5}$$

Proof. Let $L_1, L_2, \ldots, L_k = L_o$ be k points dividing Γ into k arcs of equal length F/k. Call t_i the curvilinear abscissa of L_i. The perpendiculars to Γ at L_i partition $\wedge(z_n)$ into k areas

$D_{nr}(z_n)$ with surface $A_{nr}(z_n)$, where $r = 1,\dots,k$ ($D_{n1}(z_n)$ corresponds to arc $\widehat{L_0L_1}$, we have $L_0 = M_0$, $t_0 = 0$ and $t_k = F$). Set

$$I_{nr}(z_n) = \left\{i \mid i\in[1,n],\quad X_i\in D_{nr}(z_n)\right\}$$

$$E_{nr}(z_n) = \left\{\forall\ i = 1,\dots,n,\quad X_i\notin D_{nr}(z_n)\right\}$$

$$Z_{nr} = \inf_{I_{nr}(z_n)} d(X_i,\Gamma)$$

and

$$Z_n = \sup_r Z_{nr}$$

then

$$P(Z_n < z_n) = P(\cap_r \overline{E_{nr}}(z_n)) = 1 - P(\cup_r E_{nr}(z_n)) \tag{17}$$

By construction, the family $E_{nr}(z_n)$ satisfies (A) (see [2], p. 344-345). We have

$$\begin{aligned} A_{nr}(z_n) &= z_n\ F/k - \frac{1}{2} z_n^2 \int_{t_r}^{t_{r+1}} dt/R_t \\ &= z_n\ F/k - \frac{1}{2} z_n^2 \frac{F}{k} O(1) = F\ z/n + Lk/nc - O(k(Lk)^2/n^2) \end{aligned} \tag{18}$$

O, as well as the following O's, is uniform with respect to r, because $1/|R_t| < 1/S$. Hence

$$\begin{aligned} P(E_{nr}(z_n)) &= (1 - cA_{nr}(z_n))^n \\ &= \exp(-cFz - Lk + O(k(Lk)^2/n)) \end{aligned} \tag{19}$$

and for any fixed integer ν smaller than k,

$$P\left(\bigcap_{i=1}^{\nu} E_{nr_i}(z_n)\right) = \left(1 - c\sum_{i=1}^{\nu} A_{nr_i}(z_n)\right)^n$$
$$= \exp(-\nu\, cFz - \nu\, Lk + O(k(Lk)^2/n)) \tag{20}$$

It follows that

$$P\left(\bigcap_{i=1}^{\nu} E_{nr_i}(z_n)\right)/\left(\prod_{i=1}^{\nu} P(E_{nr_i}(z_n)\right) = 1 + o(1)$$

where $o(1)$ is uniform with respect to the r_i's,

$$\max_{1\le r\le k} P(E_{nr}(z_n)) = o(1)$$

Thus

$$P(Z_n < z_n) \underset{n}{\to} \exp(-\exp(-cFz)) \tag{21}$$

Γ is made of circular arcs with radius ρ_n. Furthermore:

$$\left\{Z_n < z_n - f_n\right\} \subset \left\{\Delta(\Gamma_n,\Gamma) < z_n\right\} \tag{22}$$

where f_n, which corresponds to the arrow of the circular arc fig. 1 (the least favorable case) is equivalent to

$$f_n \sim \frac{(F/k)^2}{2\rho_n} = F^2/2k^2\rho_n$$

It follows from (12) that

$$f_n/z_n \sim cF^3\, n/2K^3Lk\, \rho_n = o(1) \tag{23}$$

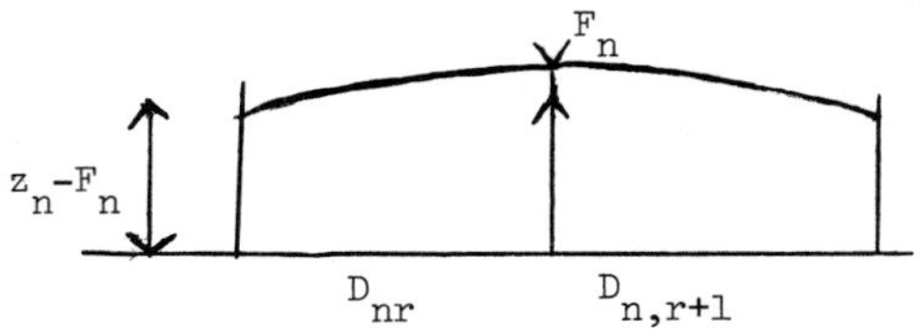

Figure 1.

and

$$(nF/k)f_n \sim nF^3/2k^3\, \rho_n = o(1) \tag{24}$$

so that

$$\begin{aligned} ncA_{nr}(z_n-f_n) &= ncz_n\, F/k + ncFf_n/k - \tfrac{1}{2}\, ncz_n^2(1-f_n/z_n)^2\, \tfrac{F}{k}\, O(1) \\ &= -cFz - Lk + o(1) \end{aligned} \tag{25}$$

o being uniform with respect to r.

Thus (19), (20) and (21) can still be written for z_n-f_n, whence (14).

From theorem 4.1. [2], p.357, we know that there exists $A > 0$ great enough to have

$$\sum(1 - P(\Gamma_n \subset \Lambda(a_n))) < +\infty \tag{26}$$

with

$$a_n = A(Ln/n)^{2/3}\, \rho_n^{-1/3} \quad . \tag{27}$$

Suppose that the event $\varepsilon = \{\Gamma_n \subset \wedge(a_n)\}$ happened. Q being a point of Γ_n let $\vec{T}_{1Q}$ and $\vec{T}_{2Q}$ be the unit vectors of the half-tangents to Γ_n at $Q(\vec{T}_{1Q} = \vec{T}_{2Q}$, except at angle points of Γ_n).

Set $Q' = v(Q)$ and call $\vec{T}_{Q'}$ the unit vector of the tangent to Γ at Q'. Let $d^1(\vec{T}_{1Q}, \vec{T}_{Q'})$ (resp. $d^1(\vec{T}_{2Q}, \vec{T}_{Q'})$) be the angle, between 0 and $\pi/2$, of the straight lines bearing $\vec{T}_{1Q}$ (resp. $\vec{T}_{2Q}$) and $\overline{T}_{Q'}$. We intend to majorize

$$U(Q) = \frac{1}{2}(d^1(\vec{T}_{1Q}, \vec{T}_{Q'}) + d^1(\vec{T}_{2Q}, \vec{T}_{Q'})).$$

According to fig. 2, the angular variation Δa_1 on Γ from B_1 to B_2 satisfies, for sufficiently great n:

$$|\Delta a_1| < 2\, B_1B_2/S$$

In the same way, for sufficiently great n, we have

$$B_1B_2^2 \leq 2a_n\rho_n = 2A(Ln/n\rho_n)^{2/3}$$

whence

$$|\Delta a_1| \leq 2\sqrt{2A}\,(Ln/n\rho_n)^{1/3}$$

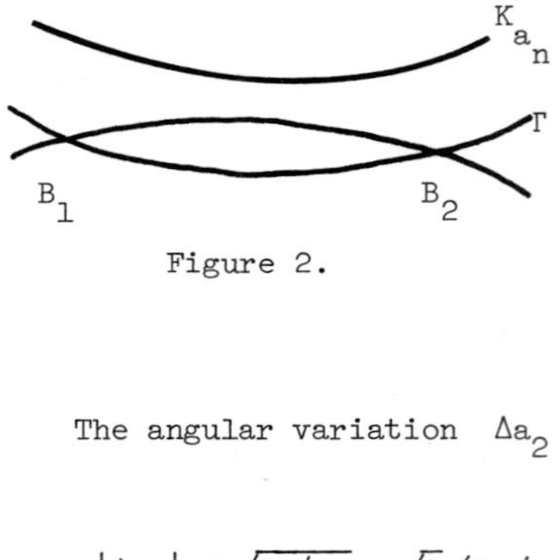

Figure 2.

The angular variation Δa_2 on the circular arc $\widehat{B_1B_2}$ satisfies

$$|\Delta a_2| \leq \sqrt{a_n/\rho_n} = \sqrt{A}\,(Ln/n\rho_n^2)^{1/3}$$

Thus, there exists $M'>0$ such that, if ε happened, then, for sufficiently great n

$$\sup_{Q\in\Gamma_n} U(Q) \leq M'\,(Ln/n\rho_n^2)^{1/3} \tag{28}$$

ϕ being the angle $(\vec{T}_o, \vec{T}_t)$, set $M^* = \int_0^F |\phi| dt$. Under the same conditions, it follows that

$$(1 - \frac{M}{2}(Ln/n\rho_n^2)^{2/3})(F - 2a_n M^*) \le F_n \le F + 2a_n M^*$$

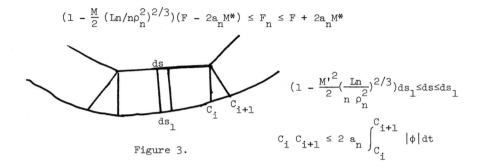

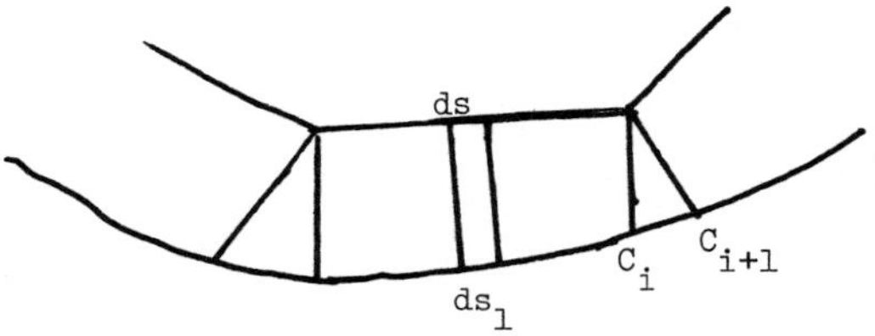

Figure 3.

$$(1 - \frac{M'^2}{2}(\frac{Ln}{n\rho_n^2})^{2/3})ds_1 \le ds \le ds_1$$

$$C_i C_{i+1} \le 2 a_n \int_{C_i}^{C_{i+1}} |\phi| dt$$

Thus there exists $M'' > 0$ such that, if ε happened, for sufficiently great n

$$|F - F_n| \le \delta_n = M''(Ln/n\,\rho_n^2)^{2/3} \tag{29}$$

Set $z''_n = \frac{kz}{n} + \frac{k\ Lk}{cn(F + \delta_n)} < z'_n$

According to (15)

$$ncz''_n = n\,c(z_n - \frac{k\ Lk \cdot \delta_n}{ncF(F - \delta_n)})$$

$$= ncz_n - \frac{k\ Lk}{F(F - \delta_n)}(\frac{Ln}{n\,\rho_n^2})^{2/3}$$

$$= ncz_n + o(1)$$

and

$$z''_n = z_n(1 + o(1)).$$

Thus we can write (19), (20) and (21) for z''_n and $z''_n - f_n$

$$P(Z_n < z''_n - f_n) \underset{n}{\rightarrow} \exp[-e^{-cFz}] \tag{30}$$

From (24), it follows that

$$P(Z_n < z'_n - f_n)/P(Z_n < z'_n - f_n/\varepsilon) \underset{n}{\rightarrow} 1 \tag{31}$$

$$P(Z_n < z'_n - f_n/\varepsilon) \geq P(Z_n < z''_n - f_n/\varepsilon) \tag{32}$$

$$P(Z_n < z''_n - f_n/\varepsilon)/P(Z_n < z''_n - f_n) \underset{n}{\rightarrow} 1 \tag{33}$$

and

$$\exp(-e^{-cF_n z}) \underset{\text{a.co.}}{\rightarrow} \exp(-e^{-cFz}) \tag{34}$$

Then (16) is a consequence of the inequality

$$P[Z_n < z'_n - f_n] \leq P[\Delta(\Gamma_n,\Gamma) < z'_n] \tag{35}$$

IV. 2nd CASE: ϕ IS LIPSCHITZIAN

In order to generalize theorem 1, we shall again use the same scheme as in the proof of [3].

Suppose ϕ lipschitzian of order δ. Suppose also that $k(n)$, apart from (11), satisfies

$$k(Lk)^{1 + 1/\delta} = o(n) \tag{36}$$

Let β_n be the only solution of equation

$$\frac{1}{F} \int_0^F \exp(Lk - \frac{n}{k} \phi(s(t)) \; F \; \beta_n) dt = 1 \tag{37}$$

Set $z_n = \frac{k}{n} z + \beta_n$ (38)

and $c = \inf_{M \in \Gamma} \phi(M)$ (39)

Lemma 1. β_n satisfies the inequalities

$$\frac{k \; Lk}{n\beta} \leq \beta_n \leq \frac{k \; Lk}{n\alpha} \tag{40}$$

Proof: Follows immediately from (3).

Lemma 2. If we set

$$\sigma_n(u) = \sum_{r=1}^{k} P(E_{nr}(u))$$

then, as n tends towards $+\infty$,

$$\sigma_n(z_n) \sim \frac{1}{F} \int_0^F \exp[Lk - \phi(s(t))(Fz + \frac{n}{k} F\beta_n)]dt \tag{41}$$

Proof:
$$A_{nr}(z_n) = \frac{F}{n} z + \frac{F}{k} \beta_n - \frac{1}{2} \frac{F}{k} z_n^2 \; O(1)$$

$$= \frac{F}{n} z + \frac{F \; \beta_n}{k} + O(\frac{k(Lk)^2}{n^2})$$

$$= \frac{F}{n} z + \frac{F \; \beta_n}{k} + \frac{\epsilon_{nr}}{n} \tag{42}$$

where ϵ_{nr}, according to (11), is a $o(1)$ uniform with respect to r.

Since ϕ is continuous, there exists $M_r \in D_{nr}(z_n)$ such that

$$P(D_{nr}(z_n)) = \left(\frac{Fz + \epsilon_{nr}}{n} + \frac{F\beta_n}{k}\right) \phi(M_r) \tag{43}$$

Set $\phi_r = \phi(M_r)$

$$W = \frac{Fz}{n} + \frac{F\beta_n}{k}$$

and $\omega_{nr} = P(E_{nr}(z_n))$ (44)

Then

$$\sigma_n(z_n) = \sum_{r=1}^{k} \omega_{nr} = \sum_{r=1}^{k} \left(1 - \left(W + \frac{\epsilon_{nr}}{k}\right)\phi_r\right)^n \tag{45}$$

consider the integral

$$\sigma'_n = \frac{1}{F} \int_0^F k(1 - W\,\phi(s(t)))^n dt \tag{46}$$

There exists $M'_r \in]\frac{r-1}{k}, \frac{r}{k}[$ $(r = 1,\dots,k)$ such that

$$\sigma'_n = \frac{1}{F} \sum_{i=1}^{k} \int_{(r-1)F/k}^{rF/k} k(1 - W\,\phi(s(t)))^n dt$$

$$= \sum_{r=1}^{k} (1 - W\,\phi(M'_r))^n \tag{47}$$

set $\phi'_r = \phi(M'_r)$

$$\omega'_{nr} = (1 - W\,\phi'_r)^n \qquad (48)$$

and $\psi_n(t) = k(1 - W\,\phi(s(t)))^n.$ (49)

According to lemma 1, we have

$$L\psi_n(t) = Lk + nL(1 - W\,\phi(s(t)))$$

$$= Lk - (Fz + \frac{nF}{k}\beta_n)\,\phi(s(t)) + O(\frac{(Lk)^2}{n})$$

$$\sigma'_n = \sum_{r=1}^{k} \omega'_{nr} \sim \frac{1}{F}\int_0^F \exp(Lk - \phi(s(t))(zF + \frac{nF}{k}\beta_n))dt \qquad (50)$$

Note that for sufficiently great n, $D_{nr}(z_n)$ is included in a rectangle of size F/k and $2\,\beta_n$.

Then the Lipschitz condition on ϕ implies

$$\phi_r = \phi'_r + \mu_{nr}((\frac{F}{k})^\delta + (\frac{k\ Lk}{n})^\delta) \qquad (51)$$

where $|\mu_{nr}|$ is majorized by a fixed real $A^* > 0$.

Thus

$$L\omega_{nr} = nL(1 - W\,\phi'_r - \frac{\epsilon_{nr}}{n}\phi_r + W\,\mu_{nr}((\frac{F}{k})^\delta + (\frac{k\ LK}{n})^\delta)) \qquad (52)$$

Using (11) and (36), it follows that

$$L\omega_{nr} = -(Fz + \frac{nF}{k}\beta_n)\phi'_r + o(1) \qquad (53)$$

where o is uniform with respect to r.

In the same way

$$L\omega'_{nr} = nL(1 - W\,\phi'_r) = -(Fz + \frac{nF}{k}\beta_n)\phi'_r + o(1) \qquad (54)$$

where o is uniform with respect to r.

As a consequence

$$\sigma_n(z_n) \sim \sigma'_n ,$$

which yields (41), by taking (50) into account.

Lemma 3. $\lim\limits_{n} \sigma_n(z_n) = \exp(-cFz)$ (55)

Proof. According to lemma 2, it is sufficient to prove that

$$\frac{1}{F}\int_0^F \exp(Lk - \phi(s(t))(Fz + \frac{nF}{k}\beta_n))dt \to e^{-cFz}$$

Set $g_n(t) = \exp(Lk - \phi(s(t))(zF + \frac{nF}{k}\beta_n))$

$$J_n(z) = \frac{1}{F}\int_0^F g_n(t)\, e^{-Fz\,\phi(s(t))}dt$$

$E_\epsilon = \{t \mid 0 \le t \le F,\ \ \phi(s(t)) < c + \epsilon\}$, with $\epsilon > 0$

$B_\epsilon = [0, F] - E_{2\epsilon}$

$$I_1 = \frac{1}{F}\int_{E_\epsilon} g_n(t)dt$$

and

$$I_2 = \frac{1}{F}\int_{B_\epsilon} g_n(t)dt$$

According to (37), we have

$$I_1 + I_2 \leq 1 \tag{56}$$

We can also write

$$I_1 \geq \exp(Lk - (c+\epsilon)\frac{nF}{k}\beta_n)\int_{E_\epsilon}\frac{dt}{F} \tag{57}$$

and

$$I_2 \leq \exp[Lk - (c + 2\epsilon)\frac{nF}{k}\beta_n]\int_{B_\epsilon}\frac{dt}{F} \tag{58}$$

which proves that $I_2/I_1 \underset{n}{\to} 0$.

From the latter property and from (56), it follows that

$$\lim_n I_2 = 0$$

and, for any $\epsilon > 0$

$$\lim_n I_1 = 1$$

As a consequence

$$J_n(z) = \frac{1}{F}\int_{E_\epsilon} g_n(t)\, e^{-Fz\,\phi(s(t))}\, dt + o(1) \tag{59}$$

Finally, suppose for instance $z > 0$. Then we have the inequalities

$$\frac{e^{-(c+\epsilon)zF}}{F}\int_{E_\epsilon} g_n(t)dt + o(1) \leq J_n(z) \leq e^{-czF}\int_{E_\epsilon} g_n(t)dt + o(1)$$

hence, for any $\epsilon > 0$

$$e^{-(c+\epsilon)Fz} \leq \underline{\lim}\, J_n(z) \leq \overline{\lim}\, J_n(z) \leq e^{-cFz}$$

which means that

$$\lim J_n(z) = \exp(-cFz) \tag{60}$$

The proof of (60) is the same for $z < 0$.

<u>Theorem 2</u>. For any $z \in \mathbb{R}$,

$$\underset{n}{\underline{\lim}}\, P[\Delta(\Gamma_n,\Gamma) < z_n] \geq \exp(-e^{-cFz}) \tag{61}$$

<u>Proof</u>: Using the same notations as in theorem 1, we shall first investigate the variable Z_n and the sequence of events $E_{nr}(z_n)$.

By construction $E_{nr}(z_n)$ satisfies (A).

According to (43), we can write

$$L(P(\bigcap_{i=1}^{\nu} E_{nr_i}(z_n))) = nL(1 - W \sum_{i=1}^{\nu} \phi_{r_i} + o(1/n))$$

$$= -nW \sum_{i=1}^{\nu} \phi_{r_i} + o(1)$$

and

$$L(\prod_{i=1}^{\nu} P(E_{nr_i}(z_n))) = \sum_{i=1}^{\nu} nL(1 - W\phi_{r_i} + o(1))$$

$$= -nW \sum_{i=1}^{\nu} \phi_{r_i} + o(1)$$

where the o's are uniform with respect to r. Hence (B).

From lemma 1, it follows that, for sufficiently great n,

$$P(E_{nr}(z_n)) < (1 - \frac{F}{2k} \beta_n \phi'_r)^n \leq (1 - \frac{\alpha F}{2\beta} \cdot \frac{k}{n})^n$$

whence (C).

As a consequence, taken lemma 3 into account

$$P[Z_n < z_n] \to \exp(-e^{-cFz}) \tag{62}$$

To reach the conclusion, take again (22) and note that (23) and (24) still hold.

REFERENCES

[1] Chevalier, J. Contribution à l'étude des statistiques d'ordre, de l'estimation géométrique et du maximum de vraisemblance. Thèse, Paris, 1975.

[2] Chevalier, J. Estimation du support et du contour du support d'une loi de probabilité. Ann. Inst. Henri Poincaré, Sect B, vol. XII, n° 4, 1976, p. 339-364.

[3] Geffroy, J. Sur un problème d'estimation géométrique. Publ. ISUP, t. 13, 1964, p. 191-200.

SOME ESTIMATION PROBLEMS FOR THE COMPOUND POISSON DISTRIBUTION

Herbert Robbins

Columbia University and
State University of New York at Stony Brook

Let (θ,X,Y) be a random vector such that

$$\text{Given } \theta,\ X \text{ and } Y \text{ are independent Poisson with mean } \theta. \tag{1}$$

Concerning the distribution function G of θ we make no assumptions. Suppose that (θ_i,X_i,Y_i), $i=1,\dots,n$, form a random sample of size n from the distribution of (θ,X,Y) and that only the "past" values X_i are observable; the parameter values θ_i and the "future" values Y_i are not. We fix some integer $a = 0,1,\dots$ and are concerned with estimating any of the three quantities

$$E_a = E(\theta|X=a) \quad \text{(an unknown constant)} \tag{2}$$

$$S = \text{the average } \theta \text{ value associated with all } X_i \text{ that equal a (a random variable)} \tag{3}$$

$$U = \text{the average } Y \text{ value associated with all } X_i \text{ that equal a (a random variable)} \tag{4}$$

To express S and U in symbols we introduce the random variables

ISBN: 0-12-166650-6

$$V_i \begin{cases} = 1 & \text{if } X_i = a \\ = 0 & \text{if } X_i \neq a \end{cases} \tag{5}$$

and

$$N_x = \text{the number of values } X_1,\ldots,X_n \text{ that equal } x \quad (x = 0,1,\ldots). \tag{6}$$

Then

$$S = \frac{\sum_1^n \theta_i V_i}{\sum_1^n V_i} = \frac{\sum_1^n \theta_i V_i}{N_a}, \quad \text{and} \tag{7}$$

$$U = \frac{\sum_1^n Y_i V_i}{\sum_1^n V_i} = \frac{\sum_1^n Y_i V_i}{N_a}. \tag{8}$$

As to E_a, since by (1)

$$f(x|\theta) = P(X=x|\theta) = e^{-\theta} \cdot \theta^x/x!, \tag{9}$$

$$f(x) = P(X=x) = \int_0^\infty f(x|\theta)dG(\theta), \tag{10}$$

we have

$$E(\theta|X=x) = \int_0^\infty \theta f(x|\theta)dG(\theta)/f(x) \tag{11}$$

$$= (x+1)f(x+1)/f(x),$$

and hence

$$E_a = (a+1)f(a+1)/f(a). \tag{12}$$

It is therefore possible to estimate E_a consistently as $n \to \infty$ by $(a+1)N_{a+1}/N_a$. To this end we define

$$W_i \begin{cases} = X_i \quad \text{if} \quad X_i = a+1 \\ \\ 0 \quad \text{if} \quad X_i \neq a+1 \end{cases} \tag{13}$$

and use

$$T = \frac{\sum_1^n W_i}{\sum_1^n V_i} = \frac{(a+1)N_{a+1}}{N_a} \tag{14}$$

to estimate E_a, S, and U. In addition to 11 we shall use the relations

$$E(\theta f(x|\theta)) = \int_0^\infty \theta f(x|\theta)dG(\theta) = (x+1)f(x+1), \tag{15}$$

$$E(\theta^2 f(x|\theta)) = \int_0^\infty \theta^2 f(x|\theta)dG(\theta) = (x+1)(x+2)f(x+2). \tag{16}$$

We begin by investigating the sampling variation of T about E_a. We have

$$\sqrt{n}(T-E_a) = \frac{1}{\sqrt{n}} \frac{\sum_1^n (W_i - E_a \cdot V_i)}{\sum_1^n V_i/n} \sim \frac{\sum_1^n (W_i - E_a \cdot V_i)}{\sqrt{n} \cdot f(a)} \quad \text{as} \quad n \to \infty. \tag{17}$$

In this sum of i.i.d. random variables the mean value of each term is

$$EW - E_a \cdot EV = (a+1)f(a+1) - \frac{(a+1)f(a+1)}{f(a)} \cdot f(a) = 0, \tag{18}$$

and the variance is

$$EW^2 + E_a^2 \cdot EV = (a+1)^2 f(a+1) + \frac{(a+1)^2 f^2(a+1)}{f(a)}. \tag{19}$$

By the central limit theorem, as $n \to \infty$

$$\sqrt{n}(T-E_a) \xrightarrow{\mathcal{D}} N(0,\sigma_1^2), \tag{20}$$

where

$$\sigma_1^2 = (a+1)^2 f(a+1)(f(a)+f(a+1))/f^3(a). \tag{21}$$

Hence

$$\frac{T - E_a}{\sqrt{(a+1)^2 N_{a+1}(N_a+N_{a+1})/N_a^3}} \xrightarrow{\mathcal{D}} N(0,1), \tag{22}$$

and an approximately 95% confidence interval for E_a is given by

$$T \pm 1.96 \sqrt{(a+1)^2 N_{a+1}(N_a+N_{a+1})/N_a^3}. \tag{23}$$

Similarly

$$\sqrt{n}\,(T-S) = \frac{1}{\sqrt{n}} \frac{\sum_1^n (W_i-\theta_i V_i)}{\sum_1^n V_i/n} \sim \frac{\sum_1^n (W_i-\theta_i V_i)}{\sqrt{n}\cdot f(a)}. \tag{24}$$

In this sum of i.i.d. random variables the mean value of each term is by (15)

$$EW - E(\theta V) = (a+1)f(a+1) - E(\theta f(a|\theta)) \tag{25}$$

$$= (a+1)f(a+1) - (a+1)f(a+1) = 0,$$

and the variance is by (16)

$$EW^2 + E(\theta^2 V) = (a+1)^2 f(a+1) + E(\theta^2 f(a|\theta)) \tag{26}$$

$$= (a+1)^2 f(a+1) + (a+1)(a+2)f(a+2),$$

so as $n \to \infty$

$$\sqrt{n}(T-S) \overset{\mathcal{D}}{\to} N(0,\sigma_2^2), \tag{27}$$

where

$$\sigma_2^2 = (a+1)((a+1)f(a+1)+(a+2)f(a+2))/f^2(a), \tag{28}$$

and

$$\frac{T - S}{\sqrt{(a+1)((a+1)N_{a+1}+(a+2)N_{a+2})/N_a^2}} \overset{\mathcal{D}}{\to} N(0,1). \tag{29}$$

Similarly, as $n \to \infty$

$$\sqrt{n}\ (T-U) \overset{\mathcal{D}}{\to} N(0,\sigma_3^2), \tag{30}$$

where

$$\sigma_2^3 = (a+1)(a+2)(f(a+1)+f(a+2))/f^2(a), \tag{31}$$

and

$$\frac{T - V}{\sqrt{(a+1)(a+2)(N_{a+1}+N_{a+2})/N_a^2}} \xrightarrow{\mathcal{D}} N(0,1). \tag{32}$$

It is easy to see that $\sigma_1^2 < \sigma_2^2 < \sigma_3^2$.

The result (32) involves only the X_i and the Y_i; it predicts the "future" from the "past" without explicit reference to the θ_i. It is interesting that (32) continues to hold when the basic hypothesis (1) is replaced by the much more general condition that for any fixed $z = 0,1,\ldots$

The conditional distribution of X, given that $X + Y = z$, is $\text{Bin}\,(z, \frac{1}{2})$. (33)

The proof of (32) under (33) proceeds directly by conditioning on the value of z. We omit the details.

It is of interest to compare the use of T to estimate E_a with a more conventional "restricted" Empirical Bayes approach in which the distribution function G of θ is assumed to be Gamma, with density function

$$g(\theta) = \frac{\alpha^\beta}{\Gamma(\beta)} \cdot e^{-\alpha\theta} \cdot \theta^{\beta-1}, \tag{34}$$

where α, β are unknown positive parameters to be estimated from $X_1,\ldots,X_n$. In this case it is easily verified that

$$E_a = (a+\beta)/(1+\alpha), \tag{35}$$

so we would replace the T of (14) by

$$\tilde{T} = (a+\tilde{\beta})/(1+\tilde{\alpha}), \tag{36}$$

where $\tilde{\alpha}$, $\tilde{\beta}$ are estimates of α, β based on $X_1,\ldots,X_n$. For example, the method of moments gives

$$\tilde{\alpha} = \frac{\bar{X}}{s^2-\bar{X}} \qquad \bar{X} = \frac{1}{n}\sum_1^n X_i,$$

where

$$\tilde{\beta} = \frac{\bar{X}^2}{s^2-\bar{X}} \qquad s^2 = \frac{1}{n}\sum_1^n (X_i-\bar{X})^2,$$

so that

$$\tilde{T} = \frac{\bar{X}^2 + (s^2-\bar{X})a}{s^2}. \tag{37}$$

If, in fact, G is Gamma, then $\tilde{T}$ will be more efficient than T, since $\sqrt{n}(\tilde{T}-E_a)$ is asymptotically normal with mean 0 and a smaller variance than (21). However, if G is not Gamma then $\tilde{T}$ need not converge to E_a at all as $n \to \infty$, in contrast to T, which always does.

REFERENCE

Robbins, H. Prediction and estimation for the compound Poisson distribution, Proc. Natl. Acad. Sci. USA 74(1977), 2670-1.

THE DECOMPOSITION OF INFINITE ORDER AND EXTREME MULTIVARIATE DISTRIBUTIONS

Paul Deheuvels

Institut de Statistique
Université Pierre et Marie Curie
Paris, France

Let $\{X_{1,n},\ldots,X_{n,n}\}$ be, $\forall\, n \geq 1$, a set of independent i.i.d. multivariate r.v., with distribution depending on n; if $X_{k,n} = (X_{k,n}(1),\ldots,X_{k,n}(p)) \in \mathbb{R}^p$, and $Y_n = \{\max_{1\leq k\leq n} X_{k,n}(1),\ldots,\max_{1\leq k\leq n} X_{k,n}(p)\}$, we say that a probability distribution P is of infinite order if there exists a sequence of Y_n whose distributions converge weakly to P. We develop here the theory of the infinite order probability distributions which includes, as a particular case the extreme values distributions. We obtain similar results as in the theory of infinite divisible distributions, a representation theorem analogous to the de Finetti theorem, and a canonical representation in terms of dependence functions, which gives, for infinite order distributions:

$$D(u_1,\ldots,u_p) = u_1 \ldots u_p \exp\left\{ \sum_{k=2}^{p} (-1)^k \sum_{1\leq i_1<\ldots<i_k\leq p} \int_0^{-\operatorname{Log} u_{i_1}} \cdots \int^{-\operatorname{Log} u_{i_k}} d\mu_{k;i_1,\ldots,i_k} \right\}$$

We study as examples the normal, Gumbel, Marshall & Olkin, Morgenstern distributions.

ISBN: 0-12-166650-6

I. THE CLASSIFICATION OF THE DEPENDENCE FUNCTION, INFINITE ORDER AND EXTREME df

A. Preliminary Results

Let $X = (X(1),\ldots,X(p))$ be a random vector of the p-dimensional real space, $F(x_1,\ldots,x_p) = P(X(1) \le x_1,\ldots,X(p) \le x_p)$ be its distribution function (pf), and $F_k(x_k) = P(X(k) \le x_k)$, $1 \le k \le p$, its marginal p.f.; we will say that $(x_1,\ldots,x_p) \in \mathbb{R}^p$ is a continuity point for the margins of F (cpm F), iff $\forall\, 1 \le k \le p$, x_k is a continuity point for F_k.

We define a <u>dependence function</u> D of F (df F), as the distribution function of a probability measure with support $[0,1]^p$, such that:

$$\forall\, 1 \le k \le p, \quad 0 \le u_k \le 1, \quad D_k(u_k) = D(1,\ldots,1, u_k,1,\ldots,1) = u_k;$$

$$\forall (x_1,\ldots,x_p) \text{ cpm } F,\ F(x_1,\ldots,x_p) = D(F_1(x_1),\ldots,F_p(x_p)). \tag{1.1}$$

We will make an extensive use of the following results, whose proofs have been detailed in [2], [3], [4].

a) Each pf F has at least one dependence function D ([2], th.(2.1), [3], th. 1), uniquely defined on the image set $(F_1(x_1),\ldots,F_p(x_p))$ of the cpm $(x_1,\ldots,x_p)$.

b) The set $\mathcal{D}$ of all dependence functions is compact with any of the following equivalent topologies: punctual convergence, uniform convergence on $[0,1]^p$, weak convergence of the associated probability measures ([2], th. 2.3).

c) If $h_1,\ldots,h_p$ are monotone, nondecreasing mappings of $\mathbb{R}$ on itself, if D is a dependence function of the pf of $(X(1),\ldots,X(p))$, it is also a df of the pf of $(h_1(X_1),\dots,h_p(X_p))$. As a by-result, if $(X(1),\dots,X(p))$ and $(Y(1),\dots,Y(p))$ are two type equivalent r.v., i.e., $\exists\ a_1,\dots,a_p > 0,\ b_1,\dots,b_p,\ (a_1X(1)+b_1,\dots,a_pX(p)+b_p) \overset{L}{\equiv} (Y(1),\dots,Y(p))$, then they have the same dependence functions. ([2], lm. (4.2)).

d) If $\{F_{(n)},\ n \geq 1\}$ is a sequence of probability distribution functions in $\mathbb{R}^p$, $F_{(n)} \to F$, where F is a probability distribution function with continuous margins, F_k, $1 \leq k \leq p$, iff:

$$\forall\ 1 \leq k \leq p,\quad F_{(n),k} \to F_k \quad \text{(punctually);}$$

If, $\forall\ n \geq 1$, D_n is a dependence function associated to $F_{(n)}$, D the dependence function associated to F, $D_n \to D$ (with the topology of $\mathcal{D}$) ([3], th. 4).

e) If $X_n = (X_n(1),\dots,X_n(p))$, $n \geq 1$, is a sequence of i.i.r.v., with pf F, and df D, if $Y_n = (\operatorname*{Max}_{1\leq i\leq n} X_i(1),\dots,\operatorname*{Max}_{1\leq i\leq n} X_i(p))$, then $D_n(u_1,\dots,u_p) = D^n(u_1^{1/n},\dots,u_p^{1/n})$ is a dependence function associated to the pf of Y_n ([2], lm. (3.1)).

We define the <u>association function</u> L of a dependence function D, by the reciprocal identities:

$$\forall\ 1 \leq i \leq p,\quad z_i \in [0, +\infty],\ L(z_1,\dots,z_p) = -\sum_{i=1}^{p} z_i - \operatorname{Log}(D(e^{-z_1},\dots,e^{-z_p}))$$

$$\forall\ 1 \leq i \leq p,\ 0 \leq u_1 \leq 1,\ D(u_1,\dots,u_p) = u_1\dots u_p \qquad (1.2)$$

$$\exp(-L(-\operatorname{Log} u_1,\dots,-\operatorname{Log} u_p)),$$

with the convention $\exp(-\infty) = 0$, $\operatorname{Log} 0 = -\infty$.

The association function (af) L_n of Y_n is thereafter:

$$L_n(z_1,\ldots,z_p) = n\ L(z_1/n,\ldots,z_p/n).$$

f) A dependence function D satisfies the following inequalities (Fréchet bounds, [2], (2.7), [3]):

$$\mathrm{Max}(0,\ 1-p+\sum_{i=1}^{p} u_i) \le D(u_1,\ldots,u_p) \le \underset{1\le i\le p}{\mathrm{Min}}\ u_i,\quad 0\le u_i\le 1,\ 1\le i\le p. \qquad (1.3)$$

Let $D_{Min}(u_1,\ldots,u_p)$ and $D_{Max}(u_1,\ldots,u_p)$ be respectively the lower and upper Fréchet bound, D_{Max} is always a dependence function, but D_{Min} is not so, if $p \ge 3$, and, in that case, it is not possible to give a lower bound as in (1.3) which is a dependence function; D_{Min} is such that:

$D_{Min}^{p-1}(u_1^{1/(p-1)},\ldots,u_p^{1/(p-1)})$ is a dependence function ([2],lm.(2.1)).

B. Order of a Dependence Function, a DE FINETTI Type Representation Theorem

We now introduce a concept which we believe to be important in the study of multivariate probability distributions:

If D is a dependence function, we define the order of D as the upper bound of the set of real numbers $r < +\infty$, such that $D^{1/r}(u_1^r,\ldots,u_p^r)$ is a dependence function. The order of D is always a number $(\sigma(D) \in [1, +\infty]$.

More generally, if F is a probability distribution function, we define the <u>order of F</u> as the upper bound of the set of the $r < +\infty$, such that $F^{1/r}(x_1,\ldots,x_p)$ is a probability distribution function. We use the same definition if $F(x_1,\ldots,x_p)$ is a monotone, nondecreasing function of each variable, such that:

$$\forall\ 1 \le i \le p, \quad \lim_{x_i \to -\infty} F(x_1,\dots,x_p) = 0, \qquad \lim_{x_1,\dots,x_p \to +\infty}$$

$F(x_1,\dots,x_p) = 1$ (in this last case, we will set $\sigma(F) = +\infty$, if $\forall\ r > 0$, $F^{1/r}$ is not a pf).

We define the set $\mathcal{D}_\infty$ of extreme dependence functions, by the fact that $D_\infty \in \mathcal{D}_\infty$ iff exists a dependence function D such that,

$$\forall\ 1 \le i \le p,\ 0 \le u_i \le 1, \quad D_\infty(u_1,\dots,u_p) = \lim_{n\uparrow\infty} D^n(u_1^{1/n},\dots,u_p^{1/n}).$$

Lemma (1.1): $D \in \mathcal{D}$ is an extreme dependence function, iff:

$$\forall\ r > 0,\ 1\le i\le p,\ 0\le u_i\le 1,\ D^r(u_1^{1/r},\dots,u_p^{1/r}) = D(u_1,\dots,u_p). \tag{1.4}$$

Proof. GALAMBOS, [8], p. 251-9, or [2], lm. (3.2).

We deduce from (1.4), that $D \in \mathcal{D}_\infty \Rightarrow \sigma(D) = +\infty$.

Proposition (1.1): If the order $R = \sigma(D)$ of $D \in \mathcal{D}$ is finite, then $D^{1/R}(u_1^R,\dots,u_p^R)$ is a dependence function.

Proof: By hypothesis, there exists a sequence $r_n \uparrow R$, such that $D^{1/r_n}(u_1^{r_n},\dots,u_p^{r_n})$ is a dependence function for $n \ge 1$. The limit of this sequence is $D^{1/R}(u_1^R,\dots,u_p^R)$, and the result is deduced by the compacity (b) of $\mathcal{D}$.

Lemma (1.2): If $D \in \mathcal{D}$ is a dependence function, $D^r(u_1^{1/r},\dots,u_p^{1/r})$ is a dependence function, if $r = 1,2,\dots,p-1$, or if r is a real number such that $r \ge p - 1$.

Proof: [2], lm. (1.3).

<u>Proposition</u> (1.2): The order $R = (\sigma(D)$ of $D \in \mathcal{D}$ is infinite iff for each $r > 0$, $D^r(u_1^{1/r},\ldots,u_p^{1/r})$ is a dependence function.

<u>Proof</u>: According to the lemma (1.2), if $r_n \uparrow +\infty$ is a sequence such that $D^{1/r_n}(u_1^{r_n},\ldots,u_p^{r_n})$ is a dependence function, $\forall\, n \geqq 1$, then, if $t \geq p-1$ is arbitrary, then $D^{t/r_n}(u_1^{r_n/t},\ldots,u_p^{r_n/t})$ is a dependence function. Thus $D^{1/r}(u_1^r,\ldots,u_p^r)$ is a dependence function for each $r > 0$.

<u>Proposition</u> (1.3): If the order $\sigma(D)$ of $D \in \mathcal{D}$ is infinite, then D satisfies the inequalities:

$$\forall\, 1 \leq i \leq p,\ 0 \leq u_i \leq 1,\quad u_1 \ldots u_p \leq D(u_1,\ldots,u_p) \leq \underset{1 \leq i \leq p}{\mathrm{Min}}\ u_i . \tag{1.5}$$

If L is the association function of D, similarly:

$$\forall\, 1 \leq i \leq p,\ 0 \leq z_i < +\infty,\quad \underset{1 \leq i \leq p}{\mathrm{Max}}\ z_i - \sum_{i=1}^{p} z_i \leq L(z_1,\ldots,z_p) \leq 0.$$

<u>Proof</u>: By the Fréchet bounds (1.3), if $e(D) = +\infty$, $\forall\, r > 0$, $\mathrm{Max}\,(0, 1-p + \sum_{i=1}^{p} u_i^r)^{1/r} \leq D(u_1,\ldots,u_p) \leq \underset{1 \leq i \leq p}{\mathrm{Min}}\ u_i$. The result is obtained by letting $r \uparrow +\infty$.

<u>Corollary</u> (1.1): 1°) If $D_1,\ldots,D_k$ are dependence functions of infinite order, $a_1,\ldots,a_k \geq 0$, $a_1+\ldots+a_k = 1$, $D_1^{a_1}\ldots D_1^{a_k}$ is a dependence function of infinite order; the set of association functions of infinite order depencence functions is convex. 2°) With the topology (b) of $\mathcal{D}$, the set of dependence functions of infinite order is compact.

<u>Proof</u>: The product of two pf is a pf; for 1°, it is thus sufficient to verify that the condition in Pr.(1.2) is satisfied. For 2°, it is likewise sufficient to verify that if $D_n \to D$, $D_n^r \to D^r$, and that D^r is a pf.

Remark: The similarities between the extreme dependence functions and the stable distributions, and between the dependence functions of infinite order and the infinitely divisible distributions are very striking. The next proposition and theorem are the versions for dependence functions of DE FINETTI's theorem (LUKACS, [12], p. 111-112).

Proposition (1.4): If D is a dependence function, $s > 0$, an arbitrary positive real number, then, the function defined by:

$$G_{s,D}(u_1,\ldots,u_p) = \begin{cases} \exp[s(D(\frac{1}{s}\text{ Log } u_1+1\ldots,\frac{1}{s}\text{ Log } u_p+1)-1], \\ \quad \text{if } \forall\ 1 \le i \le p,\ e^{-s} \le u_i \le 1, \\ \text{Min}(u_1,\ldots,u_p), \text{ if } \forall\ 1 \le i \le p,\ 0 \le u_i \le 1, \\ \quad \text{and } \underset{1\le i\le p}{\text{Min}}\ u_i \le e^{-s}, \end{cases} \tag{1.6}$$

is a dependence function of infinite order.

Proof: If $n \ge s$, D a dependence function, then the function defined by:

$$g_n(u_1,\ldots,u_p) = \begin{cases} 1 - \frac{s}{n} + \frac{s}{n} D(u_1,\ldots,u_p), \text{ if } \forall\ 1\le i\le p,\ 0\le u_i\le 1, \\ 0, \text{ if } \underset{1\le i\le p}{\text{Min}}\ u_i < 0, \end{cases} \tag{1.7}$$

is a pf of a probability measure with support in $[0,1]^p$. Since the set of the probability measures in $[0,1]^p$ is compact for the weak topology (PARTHASARATHY, [16], p.45), and since g_n^n is also a pf.

$$\underset{n\uparrow\infty}{\text{Lim}}\ g_n^n(u_1,\ldots,u_p) = \begin{cases} \exp(s(D(u_1,\ldots,u_p)-1)), \text{ if } \forall\ 1\le i\le p,\ 0\le u_i\le 1, \\ 0, \text{ if } \underset{1\le i\le p}{\text{Min}}\ u_i < 0, \end{cases} \tag{1.8}$$

is a pf. It can be verified that $G_{s,D}$ is a dependence function associated with this pf. Now we can see that $G^r_{s,D}(u_1^{1/r},\ldots,u_p^{1/r}) = G_{rs,D}(u_1,\ldots,u_p)$, $\forall\, r > 0$, and that the necessary and sufficient condition of Pr. (1.2) is satisfied by $G_{s,D}$.

<u>Theorem</u>(1.1): In order that D be a dependence function of infinite order, it is necessary and sufficient that D be a limit of the form:

$D = \lim_{n\uparrow\infty} G_{s_n,D_n}$, where $\{s_n > 0,\ n \geq 1\}$ is a sequence of positive numbers, which can be chosen such that $\lim_{n\uparrow\infty} s_n = +\infty$, or such that $s_n = n$, $\forall\, n \geq 1$, while the D_n are dependence function, $G_{s,D}$, $s > 0$, $D\in\mathcal{D}$ being defined as:

$$G_{s,D}(u_1,\ldots,u_p) = \begin{cases} \exp[s(D(\frac{1}{s}\,\mathrm{Log}\; u_1+1,\ldots,\frac{1}{s}\,\mathrm{Log}\; u_p+1)-1)] \\ \qquad \text{if } \forall\, 1 \leq i \leq p,\ e^{-s} \leq u_i \leq 1, \\ \mathrm{Min}(u_1,\ldots,u_p), \text{ if } \forall\, 1 \leq i \leq p,\ 0 \leq u_i \leq 1, \\ \qquad \text{and } \min_{1\leq i\leq p} u_i \leq e^{-s}. \end{cases} \tag{1.9}$$

<u>Proof</u>: By corollary (1.1), 2°, and Proposition (1.4), (1.9) is sufficient.

To show that it is necessary, consider:

$$H_t(u_1,\ldots,u_p) = \begin{cases} \exp(\frac{1}{t}\, D^t(u_1,\ldots,u_p)-1)), \text{ if } \forall\, 1\leq i\leq p,\ 0\leq u_i\leq 1, \\ 0 \text{ if } \min_{1\leq i\leq p} u_i \leq 0; \end{cases}$$

By (1.8), if $\sigma(D) = +\infty$, H_t is a pf of a probability distribution with support in $[0,1]^p$. Now $\lim_{t\downarrow 0} H_t(u_1,\ldots,u_p) = D(u_1,\ldots,u_p)$; according to (d), any sequence of dependence function D_n associated with $H_{1/n}$ converges to D, we may now take $D_n = G_{n,D^{1/n}} \to D$; the result follows.

Corollary (1.2): D is a dependence function of infinite order, iff exists a sequence $\{s_n, n\geq 1\}$, with $\lim_{n\uparrow\infty} s_n = +\infty$, and a sequence of dependence functions $\{D_n, n\geq 1\}$, such that the association function L of D is the limit:

$$\forall\ 1\leq i\leq p,\ 0\leq z_i < +\infty,\ L(z_1,\ldots,z_p) = -\sum_{i=1}^{p} z_i + \lim_{n\uparrow\infty} s_n \left[1 - D_n\left(1-\frac{z_1}{s_n},\ldots,1-\frac{z_p}{s_n}\right)\right] \tag{1.10}$$

Proof: Since in theorem (1.1), it is possible to choose $\{s_n, n \geq 0\}$, such that $\lim_{n\uparrow\infty} s_n = +\infty$, the result follows from (1.2) and (1.6).

We now get back to the study of general probability distribution functions:

Proposition (1.5): If F is a probability distribution function with continuous margins, the order of F is the order of its (unique) dependence function.

Proof: It follows from (c).

Theorem (1.2): For each probability distribution function F, with order $R = 1,2,\ldots\ p-1$, or $R \geq p-1$, exists a dependence function D with the same order.

Proof: By (c), we can always assume that F has a compact support $[0,1]^p$; Let R be the order of F; by a similar argument as in Proposition (1.1), we can show that $F^{1/R}$ is a pf. Now if we consider the convolution of $F^{1/R}$ by a continuous distribution weakly convergent to the Dirac distribution in 0, we can obtain a sequence $\{G_n, n\geq 1\}$ of pf with continuous margins, such that $G_n \to F^{1/R}$; we can thus extract from the sequence $\{D_n, n\geq 1\}$ of the dependence functions of the G_n, a convergent subsequence to D, which is a df of $F^{1/R}$, and such that $D^R(u_1^{1/R},\ldots,u_p^{1/R})$ is a df of F, by lemma (1.2).

Proposition (1.6): If F is a probability distribution with order R, the order of any dependence function associated to F is smaller or equal than R.

Proof: If the converse was true for D, there would be $\varepsilon > 0$, such that $D^{1/(R+\varepsilon)}$ would be a pf. Then $F^{1/(R+\varepsilon)}(x_1,\ldots,x_p) = D^{1/(R+\varepsilon)}(F_1(x_1),\ldots,F_p(x_p))$ would be a pf, which is impossible.

C. Examples

a) The Morgenstern distribution: (MARDIA, [13], p.76, MORGENSTERN, [15]) is bivariate, with dependence function $D(u,v) = uv(1+a(1-u)(1-v))$, $-1 \leq a \leq 1$; In order that D^r be a pf, it is necessary and sufficient that:

$$C_r = \frac{\partial^2 D(u,v)}{\partial u \partial v} D(u,v) - (1-r)\,\frac{\partial D(u,v)}{\partial u}\,\frac{\partial D(u,v)}{\partial v} \geq 0,\ \forall\ 0\leq u,\ v\leq 1. \quad (1.11)$$

$$\text{Here,}\quad C_r = xy\left[r\left\{(1 + a(1-x)(1-2x))(1 + a(1-y)(1-2y))\right\} + axy \right] ;$$

when $r \to 0$, $C_r \to C_0 = a\,x^2y^2$; thus, the Morgenstern distribution is of infinite order iff $0 \le a \le 1$. It can be easily verified that the order of the Morgenstern distribution is $1/(-a)$, if $-1 \le a < 0$.

This distribution gives a concrete example of a dependence function of infinite order (for $0 < a \le 1$), which is not an extreme df.

b) <u>The Marshall & Olkin distribution</u>: (MARDIA, [13], p.90, MARSHALL & OLKIN, [14]) is bivariate, with a pf: $F(x,y) = 1 - e^{-(a+c)x} - e^{-(b+c)y} + e^{-(ax+by+c\text{Max}(x,y))}$, $x,y > 0$, $a,b,c > 0$. The corresponding dependence function is:

$$D(u,v) = u + v - \text{Max}(u,v) = \text{Min}(u,v), \quad 0 \le u, \quad v \le 1.$$

This dependence function is the upper Frechet bound (1.3), $D_{\text{Max}}(u,v)$ and is an extreme value df of infinite order.

c) <u>The Gumbel bivariate logistic distribution</u>. [GUMBEL, [10], GUMBEL & MUSTAFI, [11], MARDIA, [13], p.78) has a dependence function defined for $m \ge 1$, by:

$$D(u,v) = \exp(-\{(-\text{Log } u)^m + (-\text{Log } v)^m\}^{1/m}), \quad \forall\, 0 \le u, \ v \le 1,$$

and its association function L is given by:

$$L(y,z) = (y^m + z^m)^{1/m} - y - z, \quad 0 \le y, \ z < +\infty.$$

It can be verified that D satisfies the condition (1.11), hence the distribution is of infinite order if $m \ge 1$. It can be also seen that L satisfies the (1.5) bounds, since for $m \ge 1$, $(x^m + y^m)^{1/m} \ge \text{Sup}(x,y)$, if $0 \le x, \ y < +\infty$.

d) <u>A sufficient condition for a bivariate a.c. pf to be of infinite order</u>:

<u>Theorem</u> (1.3): If F is an absolutely continuous bivariate probability distribution function, with density f, a sufficient condition

for F to be of order $R \geq 1/r$, $0 \leq r \leq 1$, is that f satisfies:

$$\forall\ x,\ y \in \mathbb{R},\ \forall\ u \leq x,\ v \leq y,\quad f(x,y)f(u,v) \geq (1-r)f(x,u)f(u,y). \tag{1.12}$$

Proof: Following (1.11), we must have, $\forall\ x,\ y \in \mathbb{R}$:

$$\int_{-\infty}^{x}\int_{-\infty}^{y} f(x,y)f(u,v)\,du\,dv \geq (1-r)\int_{-\infty}^{x}\int_{-\infty}^{y} f(x,v)f(u,y)\,du\,dv, \tag{1.13}$$

and the result follows directly.

Example: In the case a) of the Morgenstern distribution, we obtain by (1.12):

$$f(x,y)f(u,v)-f(x,u)f(u,y) = 4a(x-u)(y-v) \geq 0,\ \forall\ 0 \leq u \leq x \leq 1,\quad 0 \leq v \leq y \leq 1,$$

iff $a \geq 0$, which is a sharp result.

e) The normal bivariate distribution: We will assume that both margins are $N(0,1)$, and that the correlation coefficient is $\rho \in]-1, 1[$; the density is then:

$$f(x,y) = (2\pi)^{-1}(1-\rho^2)^{-1/2} \exp(-\tfrac{1}{2}(1-\rho^2)^{-1}(x^2-2\rho\, xy+y^2)).$$

Using (1.12), we obtain that $f(x,y)f(u,v) \geq f(x,u)f(u,y)$ iff:

$$-\rho(x-u)(y-v) \geq 0.$$

We deduce from that the fact that the normal distribution is of infinite order of $0 \leq \rho \leq 1$; reciprocally, $\forall\ x,y \in \mathbb{R}$, if $-1 < \rho < 0$, we obtain that $f(x,y)f(u,v) < f(x,u)f(u,y)$, if $u < x$, $v < y$; from (1.13) we obtain:

Theorem (1.4): The normal bivariate distribution is of infinite order iff the correlation coefficient ρ between variables is positive.

f) The Gumbel bivariate exponential distribution: (GUMBEL, [9]) The probability density function of this distribution is:

$$f(x,y) = [(1+ax)(1+ay)-a]e^{-x-y-axy}, \quad x > 0,\ y > 0,\ 0 \le a \le 1.$$

Using (1.12), we obtain that, for $0 < u < x=o(1)$, $0<v<y=o(1)$, $f(x,y)f(u,v) - f(x,v)f(u,y) = (a^2-a)(x-u)(y-v) + o(x^2+y^2)$, thus, it can be deduced of (1.13), that if $0 < a < 1$, the order of the distribution is finite. When $a = 0$, we obtain the independency case.

D. A General Lemma About Monotone Nondecreasing Multivariate Functions.

We will call pseudo probability distribution function (pseudo-pf) any function F monotone, nondecreasing of each variable, and such that:

$$\forall\ 1 \le i \le p, \quad \lim_{x_i \to -\infty} F(x_1,\ldots,x_p)=0, \qquad \lim_{x_1,\ldots,x_p \to +\infty} F(x_1,\ldots,x_p) = 1.$$

It is a well known fact that a pseudo-pf is generally not a pf, except when $p = 1$. In fact, the essential supplementary condition for F to be a pf can be expressed easily if F is C^p (BILLINGSLEY, [1], p. 226, [2], (1.9)), by:

$$\forall\ x_1,\ldots,x_p \in \mathbb{R}, \quad \frac{\partial^p F(x_1,\ldots,x_p)}{\partial x_1 \ldots \partial x_p} = f(x_1,\ldots,x_p) \ge 0.$$

We now prove the following result, which shows that in fact, a pseudo-pf is very close to a true pf:

Lemma (1.3): If F is a pseudo probability pf such that $F(x_1,\ldots,x_p) > 0$, $\forall\ 1 \le i \le p$, $\dfrac{\partial F(x_1,\ldots,x_p)}{\partial x_i} > 0$, and $\dfrac{\partial^p F(x_1,\ldots,x_p)}{\partial x_1\ldots\partial x_p}$ exists,

then $$\lim_{R\infty} \frac{\partial F^R(x_1,\ldots,x_p)}{\partial x_1\ldots\partial x_p} = +\infty.$$

Proof: $$\frac{\partial F^R}{\partial x_1\ldots\partial x_p} = R(R-1)\ldots(R-p+1)\left\{\prod_{i=1}^{p} \frac{\partial}{\partial x_i}\right\}F^{R-P}+\ldots+RF^{R-1}\frac{\partial^p F}{\partial x_1\ldots\partial x_p};$$

The limit when $R \to +\infty$ is the limit of the leading term.

The general meaning of the result is that any roughly regular pseudo-pf F can be transformed in F^R which get locally arbitrarely close to a pf for big enough R. This gives, in fact, a general procedure for establishing pf.

II. THE CANONICAL REPRESENTATION OF INFINITE ORDER AND EXTREME DEPENDENCE FUNCTIONS

A. Representation of Extreme and Infinite Order Dependence Functions

The representation of the extreme dependence function has been established in [2],[4]. We will here summarize the results, and the methods which will permit us to solve the problem of the representation of infinite order dependence functions.

We define first, if D is a dependence function, which is the distribution function of a r.v. $(U_1,\ldots,U_p)$, the co-dependence function D^* of D, as the distribution function of $(1-U_1,\ldots,1-U_p)$. It is easily seen that:

a) D^* is a dependence function, and $D^{**} = D$;

b) D and D^* are related by the reciprocal formulas:

$\forall\ 1 \le i \le p,\quad 0 \le u_i \le 1,$

$$1-D(1-u_1,\dots,1-u_p) = \sum_{1\le i\le p} D^*(1,\dots,u_i,\dots,1)$$

$$- \sum_{1\le i<j\le p} D^*(1,\dots,u_i,\dots,u_j,\dots,1)$$

$$+\dots+(-1)^{r+1} \sum_{1\le i_1\le\dots\le i_r\le p} D^*(1,\dots,u_{i_1},\dots,u_{i_r},\dots,1)$$

$$+\dots+(-1)^{p+1}D^*(u_1,\dots,u_p);$$

$$D^*(u_1,\dots,u_p) = \sum_{1\le i\le p} (1-D(1,\dots,1-u_i,\dots,1)) \tag{2.1}$$

$$- \sum_{1\le i\le j\le p} (1-D(1,\dots,1-u_i,\dots,1-u_j,\dots,1))$$

$$+\dots+(-1)^{r+1} \sum_{1\le i_1<\dots<i_r\le p} (1-D(1,\dots,1-u_{i_1},\dots,1-u_{i_r},\dots,1))$$

$$+\dots+(-1)^{p+1}(1-D(1-u_1,\dots,1-u_p)).$$

<u>Theorem</u> (2.1): If D is a dependence function, $D^n(u_1^{1/n},\dots,u_p^{1/n})$ converges to a limit $D_\infty(u_1,\dots,u_p)$ which is an extreme dependence function, with an association function L_∞, iff:

$$\forall\ 1\le i\le p,\ 0\le z_i< +\infty\ L_\infty(z_1,\dots,z_p) = -\sum_{i=1}^{p} z_i+\operatorname*{Lim}_{n\infty} n\left[1-D\left(1-\frac{z_1}{n},\dots,1-\frac{z_p}{n}\right)\right]$$

Furthermore, if there is a sequence $s_n \to +\infty$, such that the limit:

$$\forall\ 1\le i\le p,\ 0\le z_i<+\infty,\ L(z_1,\dots,z_p)=-\sum_{i=1}^{p} z_i+\operatorname*{Lim}_{n\uparrow\infty} n\left[1-D\left(1-\frac{z_1}{s_n},\dots,1-\frac{z_p}{s_n}\right)\right] \tag{2.2}$$

exists, then L is the association function of an extreme dependence function.

Proof: [2], th. 3.4.

Comparing (2.2) to (1.10), we see that the extreme dependence functions are the limits of the dependence functions of the extremes of one particular distribution, while the infinite order dependence functions are limits of the dependence functions of a mixture of extremes of distributions. This remark can be useful to investigate the generation of such distributions.

Lemma (2.1): Let $\{s_n,\ n\ge 1\}$ be a sequence of positive numbers, $\{D_n,\ n\ge 1\}$, a sequence of dependence functions, the limit $\operatorname*{Lim}_{n\uparrow\infty} s_n\left[1-D\left(1-\frac{z_1}{s_n},\dots,1-\frac{z_p}{s_n}\right)\right]=L(z_1,\dots,z_p)+\sum_{i=1}^{p} z_i$ exists

$\forall\ 1\le i\le p,\ 0\le z_i<+\infty$, iff $\forall\ 1\le k\le p,\ 1\le i_1<\dots<i_k\le p$, the limit

$$\operatorname*{Lim}_{n\uparrow\infty} s_n\ D^*\left(1,\dots,1,\frac{z_{i_1}}{s_n},\dots,\frac{z_{i_k}}{s_n},1,\dots,1\right)=L^*(+\infty,\dots,+\infty,z_{i_1},\dots,z_{i_k},+\infty,\dots,+\infty) \tag{2.3}$$

exists, $\forall\ 1\le j\le k,\ 0\le z_{i_j}<+\infty$. L^* and L being related by the reciprocal formulas:

$$\forall\ 1\le i\le p,\ 0\le z_i < +\infty,\ L(z_1,\dots,z_p) = -\sum_{1\le i<j\le p} L^*(+\infty,\dots,z_i,\dots,z_j,\dots,+\infty)$$

$$+\dots+(-1)^{k+1}\sum_{1\le i_1<\dots<i_k\le p} L^*(+\infty,\dots,z_{i_1},z_{i_k},\dots,+\infty)$$

$$+\dots+(-1)^{p+1} L^*(z_1,\dots,z_p);$$

$$\forall\ 1\le i_1<\dots<i_k\le p,\ 0\le z_{i_j} < +\infty,\ L^*(+\infty,\dots,z_{i_1},\dots,z_{i_k},\dots,+\infty)= \qquad (2.4)$$

$$\sum_{1\le j\le k} L(0,\dots,z_{i_j},\dots,0) - \sum_{1\le j<r\le k} L(0,\dots,z_{i_j},\dots,z_{i_r},\dots,0)$$

$$+\dots+(-1)^{k+1} L(0,\dots,z_{i_1},\dots,z_{i_k},\dots,0).$$

Proof: It can be obtained easily from (2.1), noting that by (1.5), if at least a z_i, $1\le i\le p$, is finite, $0 \le L^*(z_1,\dots,z_p) \le \operatorname{Min}_{1\le i\le p} z_i < +\infty$. Note that L^* is not an association function.

Lemma (2.2): $\forall\ 1\le i_1 <\dots< i_k\le p,\ 0\le z_{i_j} < +\infty,\ L^*(+\infty,\dots,z_{i_1},\dots,z_{i_k},\dots,+\infty)$ defined by (2.3) is the distribution function of a positive measure $\mu_{k;i_1,\dots,i_k}$ on

$$E_{k;i_1,\dots,i_k}=\left\{z_1,\dots,z_p;\ 0\le z_{i_j}<+\infty,\ i\le j\le k,\ z_i = 0 \text{ if } i\notin\{i_1,\dots,i_k\}\right\} \qquad (2.5)$$

Proof: Since, by (1.5), $s_n D^*_n\left[1,\dots,\frac{z_{i_1}}{s_n},\dots,\frac{z_{i_k}}{s_n},\dots,1\right] \le \operatorname{Min}_{1\le j\le k} z_{i_k}$, the sequence of positive measures associated to this sequence of distribution functions on $E_{k;i_1,\dots,i_k}$ is uniformly bounded on each compact subset of $E_{k;i_1,\ldots,i_k}$; since the set of such measures is weakly compact, the limit of the distribution functions is the distribution function of a positive measure, bounded on each compact set.

We have for L^* the following relations:

a) $\forall\ 1\le j_1<\ldots<j_{k'}\le p,\ \{i_1,\ldots,i_k\}\subset\{j_1,\ldots,j_{k'}\},\ 0\le z_{j_m}<+\infty,\ 1\le m\le k',$

$$0 \le L^*(+\infty,\ldots,z_{j_1},\ldots,z_{j_{k'}},\ldots,+\infty)\le L^*(+\infty,\ldots,z_{i_1},\ldots,z_{i_k},\ldots,+\infty)< +\infty;$$

b) $\forall\ 1\le i_1<\ldots<i_k\le p,\quad 0\le z_{i_j}<+\infty,\ 1\le j\le k,$ (2.6)

$$L^*(+\infty,\ldots,z_{i_1},\ldots,0,\ldots,z_{i_j},\ldots,+\infty) = 0;$$

c) $\forall\ 1\le i\le p,\ L^*(+\infty,\ldots,+\infty,\ z_i,\ +\infty,\ldots,+\infty) = z_i,\ 0\le z_i < +\infty.$

According to these relations, if $\partial E_{k;i_1,\ldots,i_k}$ is the subset of $E_{k;i_1,\ldots,i_k}$ of the points $(z_1,\ldots,z_p)$ such that at least one $z_{i_j} = 0$, then:

$$\mu_{k;i_1,\ldots,i_k}(\partial E_{k;i_1,\ldots,i_k}) = 0.$$

We deduce from the lemmas (2.1) and (2.2) and from the corollary (1.2) the:

<u>Theorem</u> (2.2): If D is a dependence function of infinite order, there exists a set of positive measures: $\forall\ 1\le i_1<\ldots<i_k\le p$, $\mu_{k;i_1,\ldots,i_k}$ on $E_{k;i_1,\ldots,i_k}$ defined by (2.5), such that (2.6) $\mu_{k;i_1,\ldots,i_k}(\partial E_{k;i_1,\ldots,i_k})$ $= 0$, and:

$$\forall\ 1\le j_1<\ldots<j_{k'}\le p,\ \{i_1,\ldots,i_k\}\subset\{j_1,\ldots,j_{k'}\},$$

$$\int_{\text{fixed}(z_{i_1},\ldots,z_{i_k})} d\mu_{k';j_1,\ldots,j_{k'}}(z_1,\ldots,z_p) \le d\mu_{k;i_1,\ldots,i_k}(z_1,\ldots,z_p);$$

$$\forall\ 1\le i\le p,\quad d\mu_{1;i}(z_1,\ldots,z_p) = dz_i; \tag{2.7}$$

$$\forall\ 1\le i\le p,\ 0\le z_i<+\infty,\ L(z_1,\ldots,z_p) = \sum_{k=2}^{p}(-1)^{k+1}\sum_{1\le i_1<\ldots<i_k\le p} \int_0^{z_{i_1}}\cdots\int_0^{z_{i_p}} d\mu_{k;i_1,\ldots,i_p}(z_1,\ldots,z_p);$$

$$\forall\ 1\le i\le p,\ 0\le u_i\le 1,\ D(u_1,\ldots,u_p) = u_1\ldots u_p\exp\left\{\sum_{k=2}^{p}(-1)^k\sum_{1\le i_1<\ldots<i_k\le p} \int_0^{-\mathrm{Log}\,u_{i_1}}\cdots\int_0^{-\mathrm{Log}\,u_{i_k}} d\mu_{k;i_1,\ldots,i_k}(z_1,\ldots,z_p)\right\}. \tag{2.8}$$

Remarks: 1°) A necessary and sufficient condition for D given by (2.8) to be an extreme dependence function is that $\forall\ 1\le k\le p,\ 1\le i_1<\ldots<i_k\le p$, the measures $\mu_{k;i_1,\ldots,i_k}$ be homogenous, in the sense that:

$$\forall\ z_1,\ldots,z_p\in[0,+\infty],\ \lambda>0,\ \int_0^{\lambda z_1}\cdots\int_0^{\lambda z_p} d\mu_{k;i_1,\ldots,i_k}(s_1,\ldots,s_p) = \lambda\int_0^{z_1}\cdots\int_0^{z_p} d\mu_{k;i_1,\ldots,i_k}(s_1,\ldots,s_p).$$

With some manipulation of the measures, this leads to the following representation, given in [2], [4], of the extreme dependence functions:

$$D(u_1,\dots,u_p) = u_1\dots u_p \exp\left\{\sum_{k=2}^{p} (-1)^k \sum_{1\le i_1<\dots< \ \le p} \right. \tag{2.9}$$

$$\left. \int_{(v_1,\dots,v_p)\in S_{k;i_1,\dots,i_k}} \operatorname{Sup}_{1\le j\le k} v_{i_j} \operatorname{Log} u_{i_j} \, d\mu_{k;i_1,\dots,i_k}(v_1,\dots,v_p)\right\},$$

where $S_{k;i_1,\dots,i_k} = E_{k;i_1,\dots,i_k} \cap S_p$, $S_p = \left\{(v_1,\dots,v_p),\right.$ $\left.\sum_{i=1}^{p} v_i = 1, \ \forall\ 1\le i\le p,\ v_i \ge 0\right\}$.

2°) The best, to my knowledge, results of the preceding type, are those in GALAMBOS, [8], p.255-266. The results obtained by GALAMBOS deals with the case of extreme type convergence (th. 5.3.1), and the work is done directly on distribution functions instead of dependence functions. GALAMBOS gives a representation of the form:

$$H(x_1,\dots,x_p) = \exp\left\{\sum_{k=1}^{p} (-1)^k \sum_{1\le i_1<\dots<i_k\le p} h_{k;i_1,\dots,i_k}(x_{i_1},\dots,x_{i_k})\right\},$$

if $H(x_1,\dots,x_p) = \operatorname{Lim} F^n(a_{n,1}x_1+b_{n,1},\dots,a_{n,p}x_p+b_{n,p})$, but fails to see the effective structure on the $h_{k;i_1,\dots,i_k}$ which would be here a homogeneity in terms of the marginals distributions of H, and, in fact, uneasy to write explicitely.

Furthermore, as we have pointed out in [2], [3], the convergence of the dependence function is not related to the type convergence of the coordinates, and it is useless to mix both problems.

A further discussion for the case of extreme dependence functions is done in [2] in order to obtain necessary and sufficient conditions for the representation (2.9).

B. Necessary and Sufficient Representation

Let $\mu_{k;i_1,\ldots,i_k}$, $1\le k\le p$, $i\le i_1<\ldots<i_k\le p$, be a set of arbitrary positive Borel measures, such that:

a) $\mathrm{Supp}(\mu_{k;i_1,\ldots,i_k})\subset E_{k;i_1,\ldots,i_k}$, $\mu_{k;i_1,\ldots,i_k}(\partial E_{k;i_1,\ldots,i_k})=0$;

b) $\forall\ 1\le i\le p$, $d\mu_{1,i}(z_1,\ldots,z_p) = dz_i$;

We now establish necessary and sufficient conditions for the existence of a sequence $\{D_n^*,\ n\ge 1\}$ of dependence functions such that:

$$\forall\ 1\le k\le p,\ 0\le z_{i_1},\ldots,z_{i_k} < +\infty,\ L^*(+\infty,\ldots,z_{i_1},\ldots,z_{i_k},\ldots,+\infty)$$

$$= \int_0^{z_{i_1}} \cdots \int_0^{z_{i_k}} d\mu_{k;i_1,\ldots,i_k} = \operatorname*{Lim}_{n\infty} n\, D_n^*\left(1,\ldots,\frac{z_{i_1}}{n},\ldots,\frac{z_{i_k}}{n},\ldots,1\right).$$

If $\{a_n,\ n\ge 1\}$ is a sequence such that $\operatorname*{Lim}_{n\infty} a_n = +\infty$, and $a_n = o(n)$, the problem will be solved if we can construct a sequence of dependence functions $\{D_n^*,\ n\ge 1\}$, such that, for n big enough:

$$0 \le u_{i_1},\ldots, u_{i_k} \le \frac{a_n}{n},\ D_n^*(1,\ldots,u_{i_1},\ldots,u_{i_k},\ldots,1) = \frac{1}{n} L^*(+\infty,\ldots,nu_{i_1},\ldots,nu_{i_k},\ldots,+\infty). \tag{2.10}$$

We will first construct a sequence of probability distribution functions satisfying (2.10), with $F_n = D_n^*$.

Let dF_n be defined by $\frac{1}{n}\, dL^*(nu_1,\ldots,nu_p)$ in the set $\prod_{k=1}^{p} [0,\frac{a_n}{n}]$;

Let dF_n be defined by $\frac{1}{n} d\left[L^*(+\infty, nu_2,\dots,nu_p) - L^*(a_n, nu_2,\dots,nu_p)\right] \frac{du_1}{1-\frac{a_n}{n}}$, on the set $]\frac{a_n}{n}, 1] \prod_{i=2}^{p} [0, \frac{a_n}{n}]$, and likewise on sets of the same type;

Let dF_n be defined by:

$$dF_n = \frac{1}{n} d[L^*(+\infty,\dots,+\infty, nu_{k+1},\dots,nu_p) - \sum_{i=1}^{k} L^*(+\infty,\dots,a_n,\dots,nu_{k+1},\dots)$$
$$+\dots+ (-1)^r \sum_{1\le i_1<\dots<i_r\le k} L^*(+\infty,\dots,a_n,\dots,a_n,\dots,nu_{k+1},\dots)$$
$$+\dots+ (-1)^k L^*(a_n,\dots,a_n, nu_{k+1},\dots,nu_p)] \frac{du_1\dots du_k}{(1-\frac{a_n}{n})^k},$$

on the set $\prod_{j=1}^{k}]\frac{a_n}{n}, 1] \prod_{i=k+1}^{p} [0, \frac{a_n}{n}]$, and likewise on sets of the same type;

It can be verified that this puzzle-like construction gives at the end a distribution function F_n which satisfies (2.10); the last step is given by:

$$dF_n = \left[1 - \frac{1}{n} \sum_{k=1}^{n} L^*(+\infty,\dots,a_n,\dots,+\infty) - \dots + (-1)^p L^*(a_n,\dots,a_n)\right] \frac{du_1\dots du_p}{(1-\frac{a_n}{n})^p}, \tag{2.11}$$

on the set $\prod_{j=1}^{p}]\frac{a_n}{n}, 1]$.

Now the conditions for dF_n to be positive are that:

c) $\forall\ 1\le k\le p-1,\quad 1\le i_1<\ldots<i_k\le p,$

$$d\mu_{k;i_1,\ldots,i_k} - \sum d\mu_{k+1;i_1,\ldots,i_{k+1}} + \sum d\mu_{k+2;i_1,\ldots,i_{k+2}}$$

$$+\ldots+ (-1)^{p-k}\, d\mu_{p;1,\ldots,p} \ge 0. \tag{2.12}$$

If (2.12) is satisfied, then for n big enough (2.11) defines a positive measure on $\prod_{j=1}^{p}]\frac{a_n}{n}, 1]$, thus $dF_n \ge 0$, and F_n is a pf.

Now for $0 \le u_1,\ldots,u_p \le \frac{a_n}{n}$, if $F_{n,k}$ is the kth marginal distribution of F_n, $F_{n,k}(u_k) = u_k$. By § 1a, exists a dependence function D_n^* associated to F_n, and D_n^* satisfies (2.10). We have just proved:

<u>Theorem</u> (2.3): The representation (2.8) of the dependence functions of infinite order is necessary and sufficient iff the measures $\mu_{k;i_1,\ldots,i_k}$ satisfy the positiveness and ordering conditions (2.12).

<u>Remarks</u> 1°) When $p = 2$, the ordering condition (2.7) is necessary and sufficient. When $p = 3$, must be added the following conditions:

$$d\mu_{1;1} - d\mu_{2;1,2} - d\mu_{2;1,3} + d\mu_{3;1,2,3} \ge 0$$

$$d\mu_{1;2} - d\mu_{2;1,2} - d\mu_{2;2,3} + d\mu_{3;1,2,3} \ge 0$$

$$d\mu_{1;3} - d\mu_{2;1,3} - d\mu_{2;2,3} + d\mu_{3;1,2,3} \ge 0.$$

These conditions can be expressed without modification for the corresponding measures on the simplex S_p introduced in [2], [4]. We obtain here a correction to our sufficiency condition (3.31) in [2], which is sufficient for $p = 2$, but not for $p = 3$ or more. The point we had missed is precisely the explicit construction of the puzzle proof of theorem (2.3) when $p \geq 3$; the other details are unaffected.

2°) The following proposition gives an interesting link between extreme and infinite order dependence functions:

<u>Proposition</u> (2.1): If D is a dependence function, D belongs to the attraction domain of an extreme dependence function D_∞, iff D_∞ is the limit for n of $G_{n,D}$ infinite order dependence function associated to D by (1.6).

<u>Proof</u>: It follows from (1.10), (2.2), and (1.6).

C. Representation of Infinite Order Probability Distribution Functions

<u>Theorem</u> (2.4): F is an infinite order probability distribution function iff exists a dependence function D of F with the representation (2.8), (2.12).

<u>Proof</u>: By theorem (1.2), each pf with infinite order admits at least one df of infinite order. The proof follows by theorems (2.2) and (2.3), and by the lemma:

<u>Lemma</u> (2.3): Any unidimensional probability distribution is of infinite order.

<u>Proof</u>: Any nondecreasing monotone function with limits 0 and 1 at $-\infty$ and $+\infty$ is a pf.

Remark: It must be pointed out that the order of a probability distribution function is a notion independent (with the restrictions of Th. (1.2)) of the scales on the axis, but strongly dependent of the chosen coordinates system. For instance, for a normal df (§ 1Ce) it is always possible to choose axis such that the corresponding pf is of arbitrary order. The structure is on that point quite different of the structure of infinitely divisible distributions, which is independent of the coordinates system, but dependent of the scale.

III. CONCLUSION AND FURTHER REMARKS

There has been a very extensive early work of FRECHET ([6],[7]) on probability distribution functions. The idea of the dependence function must be traced to his work. In fact, the possible use of the dependence functions for the characterization of extreme distributions has not been emphasized until recently.

We consider that the dependence functions are the real key to the study of multivariate distributions, aside from their inherent defects in behalf of the change of coordinates systems.

Some of the preceding theorems can be expressed directly in terms of the distributions functions. The fact is that their formulation in terms of dependence or association functions is infinitely richer.

For instance, one important fact of the dependence functions is that they correspond to probability distributions with a compact support. It is naturally extremely difficult to manipulate the convergence of the pf without this hypothesis.

We have pointed out that the extreme representation theorem ([2], [4]) could be expressed as a particular case of the Choquet representation theorem for the convex compact sets. The representation theorem for the infinite order dependence functions is also of that form, since the set of the corresponding association functions is convex and compact for punctual convergence.

Furthermore, the problem of the construction of dependence functions for arbitrary distributions depends highly on the definition of the dependence function, in terms of the distribution function. We are not sure that it is possible to use a definition of the dependence function as in (1.1), without using the restriction that the points in (1.1) are continuity points for the margins. It is easy to find, for instance continuity points for the pf which are not cpm and such disturbing phenomenons affect the proofs of the existence of a dependence function. We are not aware of a utilisable definition such as (1.1) before [2], [3].

For the infinite order dependence function and distributions, that we have characterized in this work, it can be easily seen that they will give a very general alternative to the extreme distributions for the study of maximas in the non i.i.d. case.

Finite order distribution functions have some interest, first to reject their use for asymptotic extremes, second, to analyze closely their structure; for instance, the lower Fréchet bound in (1.3) is not a df if $p \geq 3$.

For the extreme values results, we may add to [2], [4], the reference [5].

REFERENCES

[1] Billingsley, P. Convergence of probability measures. Wiley, 1968.

[2] Deheuvels, P. Caracterisation complète des lois extrêmes multivariées et de la convergence des types extrêmes, Publ. Inst. Statist. Univ. Paris, XXIII, 3, 1978, p. 1-36.

[3] Deheuvels, P. Propriétés d'existence et propriétés topologiques des fonctions de dépendance avec applications à la convergence des types pour les lois multivariées. C.R. Acad. Sci. Paris, Ser. A., t.288 , p. 145-8.

[4] Deheuvels, P. Determination complète du comportement asymptotique en loi des valeurs extrêmes multivariées d'un échantillon de vecteurs aléatoires indépendants, C.R. Acad. Sci. Paris, Series A., t.288, p. 217-20.

[5] Deheuvels, P. Détermination des lois limites jointes de l'ensemble des points extrêmes d'un échantillon multivarié, C. R. Acad. Sci. Paris, Ser. A., t-288, p. 631-34.

[6] Frechet, M. Sur les tableaux de corrélation dont les marges sont données, Ann. Univ. Lyon, Ser. 3, Sci., A, 1951, p.53-77.

[7] Frechet, M. Sur les tableaux de corrélation dont les marges sont données, C. R. Acad. Sci. Paris, T. 242, 1956, p. 2426-2432.

[8] Galambos, J. The asymptotic theory of extreme order statistics, Wiley, 1978.

[9] Gumbel, E.J. Bivariate exponential distributions, J. Amer, Stat. Assoc., 1960, 55, p.698-707.

[10] Gumbel, E.J. Bivariate logistic distributions, J. Amer. Stat. Assoc., 1961, 56, p. 335-49.

[11] Gumbel, E.J. and Mustafi, C.K. Some analytical properties of bivariate extremal distributions, J. Amer. Stat. Assoc., 62, p. 569-588, 1967.

[12] Lukacs, E. Characteristic functions, 2nd ed., 1969, Griffin.

[13] Mardia, K.V. Families of bivariate distributions, 1970, Griffin.

[14] Marshall, A.W. and Olkin, I. A multivariate exponential distribution, J. Amer. Stat. Assoc., 62, 1967, p. 30-44.

[15] Morgenstern, D. Einfache Beispiele zweidimensionaler Verteilungen, Mitt. Math. Statist., 8, 1956, p. 235-235.

[16] Parthasarathy, K.R. Probability measures on metric spaces, 1967, Academic Press.

CORRECTION TERMS FOR MULTINOMIAL LARGE DEVIATIONS

James Reeds

Department of Statistics
University of California
Berkeley, California

ABSTRACT

This paper gives a slightly different approach to multinomial large deviation correction terms from that found in Hoeffding, 1965. The new method pays close attention to the geometrical shape of the set whose multinomial probability content we wish to approximate. The new theory shows, in a clear way that the mysterious powers of n appearing in the correction factor to the usual "Chernoff" or entropy approximation have their genesis in the differential topology of the Kullback-Leibler function and the set in question.

NOTATION

Let Δ^k denote the probability simplex in k-space; Δ^k is a $k-1$ dimensional polyhedron. Points in Δ^k represent probability measures on a set with k elements.

Let $\Lambda_n \subseteq \mathbb{R}^k$ be the lattice of points whose coordinates are all integer multiples of $1/n$. Let Δ_n^k be $\Lambda_n \cap \Delta^k$.

Let $n X_n$ be a random k dimensional multinomial vector with expectation np for some $p \in \Delta^k$; thus $X_n \in \Delta_n^k$ is the (random) empirical measure corresponding to the measure p.

If a_n and b_n are two positive real sequences let $a_n \overset{\cup}{\cap} b_n$ be shorthand for $\log a_n - \log b_n = O(1)$.

ISBN: 0-12-166650-6

If $S \subseteq \mathbb{R}^k$ and $t \geq 0$, let $S^t = \bigcup_{x \in S} B(x,t)$ and let $S^{-t} = \{x\colon\ B(x,t) \subseteq S\}$, where $B(x,t)$ is the open ball of radius t centered at x.

The boundary of a set S is ∂S, the closure is $\overline{S}$, the complement is S^c. The context makes clear what the enveloping space is. In particular, Δ^k is thought of as a subset of the $k-1$ dimensional space $\{(v_1,\ldots,v_k)\colon\ \sum v_i = 1\}$, so $\partial\Delta^k = \{v_1,\ldots,v_k) \in \Delta^k\colon\ v_i = 0$ for some $i\}$.

I. INTRODUCTION

It is well known that if A is open in Δ^k,

$$\ell n\ P(X_n \in A) = -n\ K(A,p) + o(n) \tag{1}$$

where $-K(A,p)$ is the entropy of the set A, given by $K(A,p) = \inf_{v \in A} k(v,p)$, where $k(q,p)$ is the notorious Kullback-Leibler information number.

This paper gives a way of replacing the $o(n)$ term in formula (1) by an expression of the form $\lambda \log n + O(1)$. This idea is not new: Sanov (1957) suggested that, for a certain wide class of sets A, $\lambda = -1/2$, and in 1965 Hoeffding a) showed that Sanov's claim, as stated, was false, and b) proved that Sanov's $\lambda = -1/2$ formula was correct for a slightly more restricted class of sets.

This paper gives results very similar to Hoeffding's, but, because it uses a slightly different technique, Hoeffding's results are slightly extended.

Briefly, the present paper pays more attention to the geometrical shape of A then Hoeffding's does, and relies less on integration techniques.

All of these results - Sanov's, Hoeffding's, as well as the present ones - rely on Stirling's formula.

Let $v = (v_1,\ldots,v_k) \in \Delta_n^k$; then

$$P(X_n = v) = n! \prod_{i=1}^{k} \frac{p^{nv_i}}{(nv_i)!} = n! \prod_{i=1}^{k} \frac{nv_i + 1}{(nv_i+1)!} \; p^{nv_i}$$

$$\asymp n^{n+\frac{1}{2}} e^{-n} \prod_{i=1}^{k} \frac{e^{nv_i}}{(nv_i+1)^{nv_i}} \cdot \frac{e^{nv_i \ln p_i}}{\sqrt{nv_i + 1}}$$

$$= n^{\frac{1-k}{2}} \prod_{i=1}^{k} \frac{1}{(v_i + \frac{1}{n})^{nv_i}} \; \frac{1}{(v_i + \frac{1}{n})^{1/2}} \; e^{nv_i \ln p_i}$$

$$\asymp n^{\frac{1-k}{2}} e^{-n \sum v_i \ln \frac{v_i}{p_i}} \prod_{i=1}^{k} \frac{1}{(v_i + \frac{1}{n})^{1/2}} ,$$

uniformly in $v \in \Delta_n^k$. If none of the coordinates of v equal zero, this may be replaced by

$$P(X_n = v) \asymp n^{\frac{1-k}{2}} e^{-n \sum v_i \ln v_i/p_i}$$

$$= n^{\frac{1-k}{2}} e^{-n\, k(v,p)} .$$

It is tempting to approximate

$$P(X_n \in A) = \sum_{v \in A \cap \Delta_n^k} P(X_n = v)$$

by $$n^{\frac{1-k}{2}} \frac{\mathrm{card}(\Delta_n^k)}{\mathrm{vol}(\Delta^k)} \int_A e^{-n\, k(v,p)}\, dv$$

$$\asymp n^{\frac{k-1}{2}} \int_A e^{-n\, k(v,p)}\, dv.$$

This paper shows how such approximations are valid if the shape of A is sufficiently regular. Section II below gives a general result on approximating sums of the form $\Sigma\, e^{-n\phi(x)}$ by integrals $\int e^{-n\phi(x)}dx$. Section III gives examples suggesting that the theorem of section II cannot be much improved. Section IV discusses the approximate evaluation of such integrals, and section V, using the new method, gives an extension of a result of Hoeffding's on the distribution of the log likelihood ratio statistic for testing simple null versus general alternative for the multinomial.

II. APPROXIMATING SUMS BY INTEGRALS

Change the notation in this section a bit. Let $A \subseteq \mathbb{R}^k$ be open, let $\phi\colon \mathbb{R}^k \to \mathbb{R}$ be a Lipschitz continuous function, with Lipschitz constant $||\phi||$, so that

$$|\phi(x) - \phi(y)| \leq ||\phi|| \quad |x-y|$$

for all $x, y \in \mathbb{R}^k$. Let Λ_n be more general than above: for some sequence x_n in $\mathbb{R}^k$, let $\Lambda_n = x_n + \mathbb{Z}^k/n$

Let $S(n) = \sum_{x\in\Lambda_n\cap A} e^{-n\phi(x)}$,

let $$I(n) = n^k \int_A e^{-n\phi(x)} \; dx.$$

We want to know, when are $I(n)$ and $S(n)$ of the same order of magnitude, i.e, when is $\ln S(n) = \ln I(n) + O(1)$? It should be clear that the Euler-MacLaurin formula cannot answer this question.

The main result is, if A is a set with "minimally smooth boundary", then $I(n) \overset{\cup}{\cap} S(n)$. Roughly, this means the boundary of A is the graph of a Lipschitz function. The following series of definitions makes this precise. (The definitions and the proof of theorem 1 are taken, almost verbatim, from Stein, 1970.)

One kind of domain with minimally smooth boundary is a special Lipschitz domain:

Definition 1. An open set $D \subseteq \mathbb{R}^k$ is a special Lipschitz domain if there is a function ψ: $\mathbb{R}^{k-1} \to \mathbb{R}$ and a constant K such that

a) $|\psi(x)-\psi(y)| \leq K|x-y|$ for all $x, y \in \mathbb{R}^{k-1}$,

and b) $D = \{(x_1,\ldots,x_k)\colon\ x_k < \psi(x_1,\ldots,x_{k-1})\}$.

The smallest constant K possible in a is denoted by $K^*(D)$.

A slight generalization of this is obtained by rotating:

Definition 2. An open set $D \subseteq \mathbb{R}^k$ is a rotated Lipschitz domain if there is an orthogonal transformation σ of $\mathbb{R}^k$ such that $\mathcal{O}D$ is a special Lipschitz domain. In that case the "bound" of D is

$$K(D) = \min\{K^*(\mathcal{O}D)\colon\ \mathcal{O} \text{ orthogonal, } \mathcal{O}D \text{ special Lipschitz}\}.$$

Finally, the main definition,

Definition 3. An open set $A \subseteq \mathbb{R}^k$ is a domain with minimally smooth boundary if there exist non-negative constants N, K and ε, and a sequence of open sets $U_1, U_2, \ldots,$ in $\mathbb{R}^k$ such that:

a) For each $b \in \partial A$, there is an i such that $B(b,\varepsilon) \subset U_i$.

b) No point in $\mathbb{R}^k$ is contained in more than N of the U_i.

c) For each i there is a rotated Lipschitz domain A_i, with $K(A_i) \leq K$, such that $A \cap U_i = A_i \cap U_i$.

Let $B(A)$ be the smallest that $\max(N,K,1/\varepsilon)$ can be.

With this definition, we can state our main result:

Theorem 1. Suppose $\phi\colon \mathbb{R}^k \to \mathbb{R}$ is Lipschitz and A has minimally smooth boundary. Then $S(n) \overset{\cup}{\cap} I(n)$, uniformly in ϕ and A, as long as $B(A)$ and $||\phi||$ are bounded.

Proof. Let $\eta\colon \mathbb{R}^k \to \mathbb{R}$ be a non-negative C^∞ function supported in the unit ball, with integral one. Let $\eta_t(x) = t^{-k}\,\eta(x/t)$; η_t also has integral one, but is supported in the t-ball.

Let ε be as in Definition 3. Let $U_0 = \{x\in A\colon \ \operatorname{dist}(x,\partial A) > \varepsilon/8\}$. The sets $U_0^{-\frac{3}{4}\varepsilon}$, $U_1^{-\frac{3}{4}\varepsilon}$, ... cover not only A but also $A^{\frac{\varepsilon}{4}}$. Let χ_i be the indicator function (characteristic function) of $U_i^{-\varepsilon/2}$; let

$$\lambda_i(x) = (\eta_{\varepsilon/4} * \chi_i)\ (x).$$

It is easy to check that the functions λ_i obey the following:

1. $0 \le \lambda_i(x) \le 1$ for all x
2. λ_i is supported in $U_i^{-\varepsilon/4}$; if $x \in U_i^{-\frac{3}{4}\varepsilon}$, $\lambda_i(x) = 1$.
3. λ_i is C^∞
4. $|\nabla\lambda_i(x)|$ is bounded by C/ε,

whose C depends only on the choice of η.

For each $x \in \mathbb{R}^k$, $\lambda(x) = \sum_{i=0}^{\infty} \lambda_i(x)$ converges $\lambda(x)$ is C^∞. For each x, there are at most $N+1$ values of i for which $x \in U_i$, so $\lambda(x) \le N+1$ and $|\nabla\lambda(x)| \le (N+1)C/\varepsilon$. And if $\operatorname{dist}(x,A) < \varepsilon/4$, $\lambda_i(x) = 1$ for some i, so that then $\lambda(x) \ge 1$.

For each i define $w_i(x) = \lambda_i(x)/\lambda(x)$ if $\lambda(x) \ne 0$; define $w_i(x) = 0$ if $\lambda(x) = 0$. Then if $\operatorname{dist}(x,A) < \varepsilon/4$, $\sum_{i=0}^{\infty} w_i(x) = 1$, and for each i,

$$|\nabla w_i(x)| = |\nabla \frac{\lambda_i(x)}{\lambda(x)}|$$

$$= |\frac{\lambda_i(x)\ \nabla \lambda(x) - \lambda(x)\ \nabla \lambda_i(x)}{\lambda^2(x)}|$$

$$\leq |\lambda_i(x)|\ |\nabla\lambda(x)| + |\lambda(x)|\ |\nabla \lambda_i(x)|$$

$$\leq 2(N+1)\ C/\varepsilon.$$

Also define sets $A_i(t)$ related to the rotated Lipschitz domains A_i appearing in Definition 3, as follows. Each A_i is a rotation O_iD_i of a corresponding special domain

$$D_i = \{(x_1,\ldots,x_k): \ x_k < \psi_i(x_1,\ldots,x_{k-1})\}, \quad \text{for}$$

some Lipschitz function ψ_i.

Define $D_i(t) = \{(x_1,\ldots,x_k): \ x_k < \psi_i(x_1,\ldots,x_{k-1}) + t\}$ let $A_i(t) = O_i\ D_i(t)$. For positive values of t, $A_i(t)$ is a protruding version of A_i; for negative values of t it is a recessed version of A_i. Let $e = (0,0,\ldots,0,1) \in \mathbb{R}^k$, let $e_i = O_ie$ be the direction in which $A_i(t)$ is out thrust or recessed from A_i: $A_i(t) = A_i + te_i$. In the case $i = 0$, simply let $A_0(t) = A_0$ for all t.

Now the proof proper can begin.

Pave $\mathbb{R}^k$ with little $1/n$ -bricks: Let β_n: $\mathbb{R}^k \to \Lambda_n$ be such that $|\beta_n(x) - x| \leq |y - x|$ for all $x \in \mathbb{R}^k$, $y \in \Lambda_n$; let

$$A^{[n]} = \{x: \ \beta_n(x) \in A\}.$$

The inverse images of β_n all have the same volume v/n^k, and $|\beta_n(x) - x| \leq b/n$, where b is the radius of the inverse image of $\beta_1(0)$.

Since $n|\phi(x) - \phi(\beta_n(x))| \leq b||\phi||$, $S(n) \overset{\cup}{\cap} \tilde{S}(n)$, where $\tilde{S}(n) = n^k \int_{A^{[n]}} e^{-n\phi(x)}\, dx$.

If n is sufficiently large, $\mathrm{dist}(x,A) < \varepsilon/4$ for all $x \in A^{[n]}$, so for such n,

$$\tilde{S}(n) = \sum_{i=0}^{\infty} \int_{A^{[n]}} w_i(x)\, e^{-n\phi(x)}\, dx.$$

Let W_i be the support of w_i. It is easy to check that, for all $i > 1$,

$$A_i(-\tfrac{M}{n}) \cap W_i \subseteq A^{[n]} \cap W_i \subseteq A_i(+\tfrac{M}{n}) \cap W_i,$$

where M is chosen sufficiently large.

Thus,

$$T_-(n) \leq \tilde{S}(n) \leq T_+(n)$$

where $T_{\pm}(n) = \sum_{i=0}^{\infty} \int_{A_i(\pm \frac{M}{n})} w_i(x)\, e^{-n\phi(x)}\, dx.$

We transform variables by translating $x \to x \mp \dfrac{M e_i}{n}$, so that the integrals over $A_i(\pm \frac{M}{n})$ become integrals over A_i. The Jacobians are all unity; we apply the Lipschitz property of ϕ and the bounds on $|\nabla w_i|$ to obtain

$$\int_{A_i(\frac{M}{n})} w_i(x)\, e^{-n\phi(x)}\, dx \leq G \int_{A_i} w_i(x - \frac{M e_i}{n}) e^{-n\phi(x)}\, dx$$

$$\leq G \int_{A_i} w_i(x)\, e^{-n\phi(x)}\, dx$$

$$+ 2G \frac{M(N+1)C}{n\varepsilon} \int_{A_i \cap W_i} e^{-n\phi(x)}\, dx,$$

where $G = e^{M||\phi||}$. Similarly,

$$\int_{A_i(-\frac{M}{n})} w_i(x) e^{-n\phi(x)}\, dx$$

$$\geq \frac{1}{G} \int_{A_i} w_i(x) e^{-n\phi(x)}\, dx$$

$$- \frac{2M(N+1)C}{G\, n\, \varepsilon} \int_{A_i \cap W_i} e^{-n\phi(x)}\, dx.$$

No x is in more than $(N+1)$ of the W_i, so $\sum_{i=0}^{\infty} \int_{A_i \cap w_i} e^{-n\phi(x)}\, dx$ $\leq (N+1) \int_A e^{-n\phi(x)}\, dx$. Since $A_i \cap U_i = A \cap U_i$ it is clear that $\sum_{i=0}^{\infty} \int_{A_i} w_i(x) e^{-n\phi(x)}\, dx = \int_A e^{-n\phi(x)}\, dx$.

Hence the inequalities just derived reduce to

$$T_+(n) \leq G \left(1 + \frac{2M(1+N)^2 C}{n\varepsilon}\right) I(n)$$

and $T_-(n) \geq \frac{1}{G} \left(1 - \frac{2M(1+N)^2 C}{n\varepsilon}\right) I(n)$.

Thus $T_+(n) \overset{\cup}{\cap} T_-(n) \overset{\cup}{\cap} I(n)$, so the theorem is proven.

Note that a similar result holds for approximating sums $\sum_{x \in \Lambda_n \cap A} g(x)e^{-n\phi(x)}$ by integrals $\int_A g(x)e^{-n\phi(x)}\,dx$, as long as $|\nabla g|$ is bounded. □

III. CAN WE DO MUCH BETTER?

Theorem 1 is wasteful in that it imposes hypotheses on the whole boundary of A and on the global smoothness of ϕ. These hypotheses are not really needed except in a neighborhood of the subset of $\overline{A}$ on which ϕ is minimized. Thus, if $\inf \phi(A) < \lambda \leq \inf \phi(B)$, for $B \subseteq A$, the smooth boundary assumptions need only be imposed on $A \setminus B$, and we need assume only that $\phi|_{A \setminus B}$ is uniformly Lipschitz.

Other than this kind of trivial weakening of hypotheses, theorem 1 cannot be much improved. The following examples show that the assumption that A has a minimally smooth boundary is crucial (∂A cannot, for example, be the graph if a Hölder-continuous but not Lipschitz continuous function), and that the approximation of $I(n)$ to $S(n)$ is, in general, no more exact than theorem 1 states.

First, let $A = \{(x,y):\ 0 < x < 1,\ 0 < y < x^2\}$. The quadratic cusp at 0 prevents A from having a smooth boundary. Let $\phi(x,y) = x$. Then $S(n) = \sum_{(i,j)} e^{-i}$, where the summation extends over all i,j such that $0 < i,\ j < n,\ i^2 > nj$. Hence $S(n) = \sum_{j=1}^{n-1} \sum_{\sqrt{nj} < i < n} e^{-i} \asymp \sum_{j=1}^{n-1} e^{-\sqrt{nj}} \asymp e^{-\sqrt{n}}$. Compare this with $\overline{S}(n) =$

$$\sum_{(\frac{i}{n}, \frac{j}{n}) \in \overline{A}} e^{-i} = \sum_{j=0}^{n} \sum_{\sqrt{nj} \leq i \leq n} e^{-i} \asymp \sum_{j=0}^{n} e^{-\sqrt{jn}} \asymp 1,$$

and with

$$I(n) = n^2 \int_0^1 \int_0^{X^2} e^{-nx}\, dydx = n^2 \int_0^1 x^2\, e^{-nx} dx = \frac{1}{n} \int_0^n u^2 e^{-u} du \sim \frac{\Gamma(3)}{n} .$$

If A has smooth boundary but ϕ fails to be Lipschitz the same kind of bad approximation holds. Let $A = (0,1)$, let $\phi(x) = \sqrt{x}$. Then $S(n) = \sum_{i=1}^{n-1} e^{-n\sqrt{i/n}} = \sum_{i=1}^{n-1} e^{-\sqrt{ni}} \asymp e^{-\sqrt{n}}$. If we sum over lattice points in $\overline{A}$ instead of A, we get $\overline{S}(n) = \sum_{i=0}^{n} e^{-n\sqrt{i/n}} =$ $\sum_{i=0}^{n} e^{-\sqrt{ni}} \asymp 1$. But

$$I(n) = n \int_0^1 e^{-n\sqrt{x}}\, dx = 2n \int_0^1 e^{-nu}\, u\, du = \frac{2}{n} \int_0^n e^{-v}\, dv \sim 2/n .$$

Finally, the approximation $I(n) \asymp S(n)$ is no more accurate than the $\asymp$ symbol implies. Let $A = (\theta,1)$, for some $\theta \in (0,1)$; let $\phi(x) = x$. Then

$$S(n) = \sum_{n\theta<i<n} e^{-i} = \sum_{n\theta<i<\infty} e^{-i} + O(e^{-n}).$$

But $\sum_{n\theta<i<\infty} e^{-i} = \sum_{j=0}^{\infty} e^{-j-1-[n\theta]}$, so $S(n) = \frac{1}{e-1} e^{-[n\theta]} + O(e^{-n})$.

On the other hand, $I(n) = n \int_\theta^1 e^{-nx}\, dx = e^{-n\theta} - e^{-n}$. Hence

$\overline{\lim} \frac{I(n)}{S(n)} = e-1$ and $\underline{\lim} \frac{I(n)}{S(n)} = 1 - \frac{1}{e}$. This kind of oscillation was noticed by Bahadur and Rao, 1960.

The problem of describing the oscillation is always a number theory problem: an exercise in diophantine approximation. In the example just given, the number theory is simply the study of the sequence of the fractional parts of $n\theta$. (See Slater, 1967 who uses continued fraction methods to get a complete solution.) But in the general $\mathbb{R}^k$ case the number theory becomes much more difficult, involving the "geometry of numbers" of the set A and multiple diophantine approximation. It is unlikely that any simple "closed form" expression $w(n)$ can be given so that

$$\lim_{n\to\infty} \frac{I(n)\, w(n)}{S(n)} = 1$$

for any range of interesting sets A.

IV. ASYMPTOTIC EVALUATION OF EXPONENTIAL INTEGRALS

How do you work out an approximation to $\int_A e^{-n\phi(x)}\, dx$?

Laplace gave the definitive answer to this question. In the $\mathbb{R}^1$ case two examples make the subject clear. Suppose $\phi(x)$ is highly differentiable and takes its minimum value on $[0,1]$ at the point 0 and that $\phi'(0) > 0$. Then the integral $I(n) = \int_0^1 e^{-n\phi(x)}\, dx$ is essentially the same as $\int_0^1 e^{-n(\phi(0)+\phi'(0)x)}\, dx = e^{-n\phi(o)} \int_0^1 e^{-n\phi'(0)x}\, dx$, which, in turn, is essentially the same as $e^{-n\phi(0)} \int_0^\infty e^{-n\,\phi'(0)x}\, dx = \dfrac{e^{-n\phi(0)}}{n\,\phi'(0)}$. In another case $\phi(x)$ is again minimized at $x = 0$, but $\phi'(0) = 0$, $\phi''(0) > 0$. Then $I(n)$ is essentially the same as $\int_0^1 e^{-n(\phi(0) + \phi''(0)\frac{x^2}{2})}\, dx$ and also as $\int_0^\infty e^{-n(\phi(0) + \phi''(0)\frac{x^2}{2})}\, dx =$

$$\sqrt{\frac{\pi}{2n\phi''(0)}} \; e^{-n\phi(0)}$$

In both cases we get $e^{-n\phi(0)}$; this is inevitable by the general large deviation theory. But depending on the differential topology of ϕ at its minimum (is it a "calculus minimum" or a "linear programming minimum") we get a factor of $1/n$ or of $1/\sqrt{n}$.

In the $\mathbb{R}^k$ case the same kind of thing happens. The infimum of ϕ on A gives the exponential factor; the differential topological description of that infimum multiplies the exponential by a power of n. The topology tells what the power is.

Here are two $\mathbb{R}^k$ cases corresponding to the two $\mathbb{R}^1$ cases given above. Assume, in both cases, to avoid technicalities, that A is a C^∞ manifold with boundary, and that ϕ is C^∞. According to the "collaring theorem" (Hirsch, 1976, p.113) there is a C^∞ imbedding of $\partial A \times [0, \infty)$ into A, so ∂A has a neighborhood in A diffeomorphic to $\partial A \times [0,1)$.

In the first example assume that the infimum of ϕ on A is attained at $b \in \partial A$, that $(\nabla\phi)(b) = 0$, and that $\psi = \phi|\partial A$ has a regular critical point at b, that is, $\nabla\psi$ vanishes at b but $\nabla^2\psi$ has full rank. (This is the case Sanov mishandled in 1957: he omitted to assume ψ has a regular critical point at b.)

The strategy is to maneuver $I(n)$ into the form

$$n^k e^{-n\phi(0)} \int_0^\infty \int_{-\infty}^\infty \cdots \int_{-\infty}^\infty e^{-nx_1 - \frac{n}{2}(x_2^2 + \cdots + x_k^2)} \, dx_1 \cdots dx_k$$

$$= \frac{1}{n} \left(\frac{1}{2\pi n}\right)^{\frac{k-1}{2}} \;\substack{\cup \\ \cap}\; n^{-\frac{1+k}{2}} \quad .$$

It might be surprising that this can be done without overt use of Taylor's theorem. First, we can localize to any neighborhood V of b in A: $I(n) \asymp n^k \int_V e^{-n\phi(x)}\,dx$. By the Morse lemma (Hirsch, p.144), there is a neighborhood $V \subseteq A$ and a set of C^∞ coordinates $y_2,\ldots,y_k$ for $\partial A \cap V$ such that $\psi(x) = \phi(b) + \frac{1}{2}(y_2^2(x) + \ldots + y_k^2(x))$ for $x \in \partial A \cap V$, with $y_i(b) = 0$, where $\psi = \phi|_{\partial A}$. Extend the y_i in a C^∞ way to V. Let $\tilde{\psi}(x) = \phi(b) + \frac{1}{2}(y_2^2(x) + \ldots + y_k^2(x))$ on all of V; $\tilde{\psi}(x) = \phi(x)$ on ∂A. Then it is easy to see that $(y_1(x) = \tilde{\psi}(x) - \phi(b),\ y_2(x),\ldots,y_k(x))$ form a system of C^∞ coordinates on some sub neighborhood $\tilde{V}$ of V, and that

$$W = \left\{x:\ 0 < y_1 < \varepsilon,\ -\varepsilon < y_2 < \varepsilon, \ldots,\ -\varepsilon < y_k < \varepsilon\right\} \subseteq \tilde{V}$$

for some $\varepsilon > 0$ is a neighborhood of b.

Clearly $I(n) \asymp \tilde{I}(n) = \int_W e^{-n\phi(x)}\,dx$. Change variables in $\tilde{I}(n)$:

$$\tilde{I}(n) = e^{-n\phi(b)} \int_0^\varepsilon \int_{-\varepsilon}^\varepsilon \cdots \int_{-\varepsilon}^\varepsilon e^{-ny_1}\, e^{-\frac{n}{2}(y_2^2 + \ldots + y_k^2)}\, J(y)\, dy_1 \ldots dy_n$$

where J is the Jacobian of the transformation. Substitute $ny_1 = u_1$, $\sqrt{n}\, y_i = u_i$, $i \geq 2$, to get

$$\tilde{I}(n)\, e^{n\phi(b)}\, n\, n^{\frac{k-1}{2}} = \int_0^{n\varepsilon} \int_{-\sqrt{n}\varepsilon}^{+\sqrt{n}\varepsilon} \cdots \int_{-\sqrt{n}\varepsilon}^{+\sqrt{n}\varepsilon} e^{-u_1 - \frac{1}{2}(u_2^2 + \ldots + u_k^2)}\, J\left(\frac{u_1}{n}, \frac{u_2}{\sqrt{n}}, \ldots, \frac{u_k}{\sqrt{n}}\right) du_1 \ldots du_k$$

General weak convergence theory shows that the right hand side has limit $(2\pi)^{\frac{k-1}{2}} J(0)$. Thus $I(n) \asymp n^{-\frac{k+1}{2}} e^{-n\phi(b)}$.

The other example is where $\nabla\phi$ never vanishes, where $A = \{x: c \le \phi(x) \le d\}$ is a compact manifold with boundary. The collaring theorem guarantees the existence of a constant $\delta > c$ and coordinates (ϕ,β) so that $\tilde{A} = \{x: c \le \phi(x) < \delta\}$ is parameterized by $c \le \phi \le \delta$ and $\beta \in \{x: \phi(x) = c\}$. Then

$$\tilde{I}(n) = \int_{\tilde{A}} e^{-n\phi(x)}$$

$$= \int_c^\delta e^{-n\phi} \int_{\partial A} J(\phi,\beta)\, d\beta \;\; d\phi,$$

where J is the Jacobian. Let $\phi = c + \frac{y}{n}$;

$$\tilde{I}_{(n)} = \frac{e^{-nc}}{n} \int_0^{(\delta-c)n} e^{-u} \int_{\partial A} J(c + \tfrac{y}{n}, \beta)\, d\beta\, dy.$$

The integral clearly converges to $\int_{\partial A} J(0,\beta) d\beta$, a finite constant. As before, general entropy lore guarantees that $I(n) \asymp \tilde{I}(n)$; thus $I(n) \asymp \frac{e^{-nc}}{n}$.

Two applications to multinomial large deviations are obvious:

Application 1. Let $A \subseteq \Delta^k$ be a C^∞ manifold with boundary, $p \notin A$, let the Kullback-Leibler function $k(v,p)$ attain a unique minimum on A at $r \in \partial A$; suppose $k(v,p)|_{\partial A}$ has a nondegenerate critical point at r. If r is not in $\partial\Delta^k$, then Stirling's formula and theorem 1 show that

$$P(X_n \in A) \asymp n^{\frac{k-1}{2}} \int_A e^{-nk(v,p)}\, dv.$$

The results of this section show that the integral is $\asymp e^{-nk(r,p)}\, n^{-k/2}$, so

$$P(X_n \in A) \asymp n^{-1/2}\, e^{-nk(r,p)} = n^{-1/2}\, e^{-nK(A,p)}$$

Application 2. Let $A(\lambda) = \{v \in \Delta^k : k(v,p) \geq \lambda\}$; assume $\lambda > 0$. If $\partial A(\lambda)$ is bounded away from $\partial \Delta^k$ (i.e., in Hoeffding's notation, $\lambda < \lambda_{crit}$) there is a $\lambda' > \lambda$ such that $\partial A(\lambda')$ is also bounded away from $\partial \Delta^k$.

Then $P(X_n \in A(\lambda)) \asymp P(X_n \in A(\lambda) \setminus A(\lambda'))$, $k(v,p)$ is uniformly Lipschitz on $A(\lambda) / A(\lambda')$, and $A(\lambda) / A(\lambda')$ is a C^∞ manifold with boundary. So

$$P(X_n \in A(\lambda)) \asymp n^{\frac{k-1}{2}} \int_{A(\lambda)\setminus A(\lambda')} e^{-nk(v,p)}\, dv$$

$$= n^{\frac{k-1}{2}} \frac{e^{-nK(A,p)}}{n} = n^{\frac{k-3}{2}} e^{-nK(A,p)} .$$

The removal of the restriction on λ is described in the next section.

V. NEW RESULT FOR LOG LIKELIHOOD TEST

Let $T_n = k(X_n,p)$ be the sample entropy. If $\lambda < k(\partial\Delta^k,p)$ then the last application of the preceeding section shows that $P(T_n > \lambda) \asymp \exp(-n\lambda)\, n^{\frac{k-3}{2}}$. But what if $\lambda \geq k(\partial\Delta^k,p)$, i.e., what if

$\partial A(\lambda) \cap \partial\Delta^k \neq \phi$? Theorem 1 does not apply directly for two reasons: the entropy function $k(v,p)$ fails to be Lipschitz continuous near $\partial\Delta^k$, and $A(\lambda)$ is not minimally smooth: the surface $\{x: k(v,p)= \lambda\}$ cuts $\partial\Delta^k$ tangentially to form a logarithmic cusp.

Nonetheless, a trick lets the present geometrical method work even when $\lambda \geq k(\partial\Delta^k, p)$, and in such cases too,

$$P(T_n > \lambda) \overset{\cup}{\cap} \exp(-n\lambda)\, n^{\frac{k-3}{2}} . \tag{2}$$

For given $\varepsilon > 0$ to be chosen below define subsets of Δ^k corresponding to subsets of $\{1,2,\ldots,k\}$ as follows: to $S = \{\epsilon_1,\ldots,\epsilon_r\} \subseteq \{1,2,\ldots,k\}$ corresponds

$$\Delta_S = \{(v_1,\ldots,v_k) \in \Delta^k: \quad \begin{array}{ll} v_i < 2\varepsilon & \text{if } i \in S, \\ v_i > \varepsilon & \text{if } i \notin S. \end{array}$$

Let A_S denote $A(\lambda) \cap \Delta_S$. We will show that

$$P(X_n \epsilon A_S) \overset{\cup}{\cap} e^{-n\lambda}\, n^{\frac{k-3}{2}} \tag{3}$$

whenever A_S is non-empty. This will clearly imply (2).

If $S = \{i_1,i_2,\ldots,i_r\}$ let $A_S(u_1,\ldots,u_r) = \{(v_1,\ldots,v_k) \epsilon A_S$: $v_{i_j} = u_j$, $j = 1,\ldots,r\}$. We can choose ε so small that for each S, either $A_S = \phi$ or, whenever

$$0 \leq u_1 < 2\varepsilon,\ldots, 0 \leq u_r < 2\varepsilon, \quad A_S(u_1,\ldots,u_r) \neq \phi.$$

It is easiest to check (3) in the case when $S = \{1,2,\ldots,r\}$; all other cases can be reduced to this one by relabeling coordinates.

Now,

$$P(X_n \in A_S) = \sum_{x_1=0}^{2u\varepsilon} \cdots \sum_{x_r=0}^{2u\varepsilon} P\left\{X_n \in A_S\left(\frac{x_1}{n},\ldots,\frac{x_r}{n}\right)\right\} .$$

By applying Stirling's formula, we see that for fixed $(x_1,\ldots,x_r)$, the summand is of the same order of magnitude as

$$n^{\frac{1-k}{2}}\, n^{\frac{r}{2}} \prod_{i=1}^{r} \frac{1}{\sqrt{x_i+1}}\; e^{-n \sum_{i=1}^{r} \frac{x_i}{n} \log \frac{x_i}{np_i}}\; Q_n(A_S;\, x_1,\ldots,x_r)$$

where

$$Q_n(A_S;\, x_1,\ldots,x_r) = \sum e^{-n\, \tilde{\phi}\left(\frac{x_{r+1}}{n},\ldots,\frac{x_k}{n}\right)}$$

where the last summation extends over all $(x_{r+1},\ldots,x_k)$ such that

$$\frac{1}{n}(x_1,\ldots,x_k) \in A_S\left(\frac{x_1}{n},\ldots,\frac{x_r}{n}\right) \cap \Lambda_n,$$

and where $\tilde{\phi}(v_{r+1},\ldots,v_k) = \sum_{i=r+1}^{k} v_i\, \ell n \frac{v_i}{p_i}$.

Now $\tilde{\phi}$ is uniformly Lipschitz on $A_S(\frac{x_1}{n},\ldots,\frac{x_r}{n})$, which is a domain with minimally smooth boundary, so Theorem 1 applies. According to the methods of the previous section, then

$$Q_n(A_S:\, x_1,\ldots,x_r) \asymp n^{k-1-r}\, e^{-n\lambda + n \sum_{i=1}^{r} \frac{x_i}{n} \ell n \frac{x_i}{np_i}}\; \frac{1}{n} .$$

Thus

$$P(X_n \in A_S) \asymp n^{\frac{1-k}{2}} e^{-n\lambda} n^{k-1-r} n^{-\frac{r}{2}} \sum_{x_1=0}^{2n\varepsilon} \cdots \sum_{x_r=0}^{2n\varepsilon} \prod_{i=1}^{r} \frac{1}{\sqrt{x_i+1}}$$

$$= n^{\frac{k-3}{2}} \left(\frac{1}{\sqrt{n}} \sum_{x=0}^{2n\varepsilon} \frac{1}{\sqrt{x+1}} \right) e^{-n\lambda}$$

$$\asymp n^{\frac{k-3}{2}} e^{-n\lambda},$$ as promised.

REFERENCES

Bahadur, R.R. and Rao, R. Ranga. "On deviations of the sample mean," *Ann. Math. Statist.* *31* (1960), 1015-1027.

Hirsch, Morris W. Differential Topology. Springer, New York, 1976.

Hoeffding, W. "On probabilities of large deviations," *Proc. Fifth Berkeley Symp. Math. Statist. Prob.* *1* (1965), 203-219.

Sanov, I.N. "On the probabilities of large deviations of random variables," *Selected Translations in Mathematical Statistics and Probability* *1*, American Mathematical Society, Providence, R.I., 1961, 213-244.

Slater, N.B. "Gaps and steps for the sequence nθ mod 1." *Proc. Camb. Phil. Soc*. *63* (1967), 1115-1123.

Stein, Elias M. Singular Integrals and Differentiability Properties of Functions. Princeton University Press, Princeton, 1970.

ON A THEOREM OF HOEFFDING

P. J. Bickel[1]

Department of Statistics
University of California
Berkeley, California

W. R. van Zwet[2]

Department of Mathematics
University of Leiden
Leiden, The Netherlands

We generalize a theorem of Hoeffding's on the dispersion of the sum of the number of successes in Poisson binomial trials to Poisson multinomial trials. Using this result we study for which measures of dispersion $\Sigma_{i=1}^n X_i$ is always more dispersed when the X_i are i.i.d. with distribution $\frac{1}{n}\Sigma_i F_i$ than when the X_i are independent, $X_i \sim F_i$, $i = 1,\ldots,n$. Some curious connections with totally positive functions and infinitely divisible laws appear.

I. INTRODUCTION AND BASIC CHARACTERIZATION

Let S be the number of successes in n independent (Poisson-binomial) trials and let p_i denote the probability of success in the i-th trial, $i = 1,\ldots,n$. Let $p = (p_1,\ldots,p_n)$. If g is a real-valued function on $\{0,\ldots,n\}$, we write $E_p\, g(S)$ for the expectation of $g(S)$ under this model. In 1956 Hoeffding proved (inter alia) the following result (theorem 3 of [7]):

---

[1]This research was partially supported by the U.S. Office of Naval Research, Contract No. N00014-75-C-0444/ NR 042-036, the National Science Foundation, Grant No. MCS76 10238 A01, and by the Netherlands' Organization for Pure Scientific Research.

[2]Research supported by the Office of Naval Research, Contract No. N00014-75-C-0444/NR 042-036.

ISBN: 0-12-166650-6

Theorem. If g is strictly convex, then

$$E_p\, g(S) \le E_{\bar{p}}\, g(S) \tag{1.1}$$

where $\bar{p} = (p_.,\ldots,p_.)$ and $p_. = n^{-1}\sum_{i=1}^n p_i$. Strict inequality holds in (1.1) unless $p = \bar{p}$.

Of course the theorem implies that (1.1) holds for all convex g. We can reformulate this result, ponderously, as follows. Let X_i, $i = 1,\ldots,n$, be independent random variables each taking on values in $A = \{0,1\}$. Denote the probability distribution of X_i by F_i, let $F = (F_1,\ldots,F_n)$ and

$$\bar{F} = (F_.,\ldots,F_.),\quad F_. = \frac{1}{n}\sum_{i=1}^n F_i. \tag{1.2}$$

If $S_n = \sum_{i=1}^n X_i$ and g is a convex function on $A \oplus \ldots \oplus A$, then (with obvious notation)

$$E_F\, g(S_n) \le E_{\bar{F}}\, g(S_n). \tag{1.3}$$

In this note we drop the restriction that $A = \{0,1\}$ and address the question for what functions g does inequality (1.3) hold for all F as above and all $A \subset \mathbb{R}^m$ of a given finite cardinality. We find that it is easy to give a general characterization of these functions. Further study, however, reveals the somewhat disappointing fact that with increasing cardinality the class of functions becomes more and more restricted and that only very smooth rather special functions are left in the limit. We rely heavily on some ideas of Horn [8] and Berg, Christensen and Ressel [2], even though we shall not make use of their results explicitly.

Specifically, let

$$\mathcal{A}_k = \{A \subset \mathbb{R}^m: \quad \text{card}(A) = k\},$$

$$\mathcal{F}(A) = \{F : F_i(A) = 1, \quad i = 1,\ldots,n\},$$

$$\mathcal{F}_k = \cup\{\mathcal{F}(A) : A \in \mathcal{A}_k\},$$

$$g : \mathbb{R}^m \to \mathbb{R}^1,$$

and suppose that g is measurable. Following Donoghue [4], we say that a symmetric $k \times k$ matrix $||a_{j,j'}||$ is almost positive if and only if $\sum_{j,j'} a_{j,j'} \omega_j\omega_{j'} \geq 0$ for all $(\omega_1,\ldots,\omega_k)$ with $\sum_j \omega_j = 0$. We shall say it is strictly almost positive if strict inequality holds unless $\omega_j = 0$ for $j = 1,\ldots,k$. Similarly we shall call a matrix positive if it is positive semi-definite and strictly positive if it is positive definite in the more usual terminology.

<u>Theorem 1.1</u>. The following three assertions are equivalent for a given g and k.

(i) Inequality (1.3) holds for all n and $F \in \mathcal{F}_k$.

(ii) Inequality (1.3) holds for $n = 2$ and all $F \in \mathcal{F}_k$.

(iii) The $k \times k$ matrix $||g(x_j + x_{j'})||$ is almost positive for all $x_1,\ldots,x_k \in \mathbb{R}^m$.

Moreover, if $||g(x_j + x_{j'})||_{k\times k}$ is strictly almost positive for all sets of distinct $x_1,\ldots,x_k \in \mathbb{R}^m$, then inequality (1.3) is strict unless $F = \overline{F}$.

Note that for $k = 2$ and measurable g, (strict) almost positivity of $||g(x_j + x_{j'})||$ and (strict) convexity of g are equivalent. Thus for $k = 2$ and $m = 1$, theorem 1.1 is the same as Hoeffding's assertion with the additional information that if (1.3) is to hold for $n = 2$ and for all two-point sets $A \subset \mathbb{R}^1$ rather than just $\{0,1\}$, convexity of g on $\mathbb{R}^1$ is necessary as well as sufficient.

The main part of the proof of theorem 1.1 is a generalization of Hoeffding's theorem to "Poisson-multinomial" trials, which we shall formulate and prove as a separate lemma. Let $p = ||p_{i,j}||_{n\times k}$, where $p_{i,j} \geq 0$ for all i,j and $\sum_{j=1}^{k} p_{i,j} = 1$ for all i. Define

$$R_{k,n} = \{t = (t_1,\ldots,t_k) : t_j \text{ natural numbers, } \sum_{j=1}^{k} t_j = n\}$$

and let h be a real-valued function on $R_{k,n}$. Consider n independent trials each having possible outcomes $1,\ldots,k$ and suppose that the probability that the i-th trial has outcome j is $p_{i,j}$. Let T_j be the number of times outcome j occurs in the n trials, let $T = (T_1,\ldots,T_k)$ and write $E_p\, h(T)$ for the expectation of $h(T)$ under this model. Define the $n \times k$ matrix

$$\bar{p} = \left\| \begin{matrix} p_{.1} & \cdots & p_{.k} \\ \vdots & & \vdots \\ p_{.1} & \cdots & p_{.k} \end{matrix} \right\| \qquad \text{with } p_{.j} = \frac{1}{n} \sum_{i=1}^{n} p_{i,j}$$

which has n identical rows equal to the average of the n rows of p. Finally, let $e_1,\ldots,e_k$ denote the usual basis vectors in $\mathbb{R}^k$, thus $e_j = (\delta_{j,1},\ldots,\delta_{j,k})$ where $\delta_{j,j'}$ is the Kronecker symbol.

<u>Lemma 1.1</u>. If $||h(t + e_j + e_{j'})||_{k\times k}$ is almost positive for all $t \in R_{k,n-2}$, then

$$E_p\, h(T) \leq E_{\bar{p}}\, h(T). \tag{1.4}$$

Strict inequality holds in (1.4) if $||h(t + e_j + e_{j'})||_{k\times k}$ is strictly almost positive for all $t \in R_{k,n-2}$ and $p \neq \bar{p}$.

Proof. Let P be the set of $n \times k$ matrices $||\pi_{i,j}||$ with $\pi_{i,j} \geq 0$ for all i and j, $\sum_j \pi_{i,j} = 1$ for all i and $\sum_i \pi_{i,j} = n\, p_{.j}$ for all j. Since $E_\pi\, h(T)$ is a continuous function of $\pi \in P$, it achieves its maximum on P.

We first prove the second part of the lemma by showing that in this case $E_\pi\, h(T)$ assumes its maximum on P and $\bar{p}$ only. Let $\pi \in P$, $\pi \neq \bar{p}$ so that π contains two unequal rows. Without loss of generality suppose these are the first two rows and define $\tilde{\pi} \in P$ by

$$\tilde{\pi}_{i,j} = \tfrac{1}{2}(\pi_{1,j} + \pi_{2,j}) \qquad \text{for } i = 1,2$$

$$= \pi_{i,j} \qquad \text{for } i = 3,\ldots,n,$$

i.e. $\tilde{\pi}$ is obtained from π by replacing the first two rows by their average. Let X_i denote the outcome of the i-th trial and define

$$a_{j,j'}(\pi) = E_\pi(h(T)|X_1=j,\ X_2=j').$$

Because $||h(t + e_j + e_{j'})||$ is strictly almost positive for $t \in R_{k,n-2}$, $||a_{j,j'}(\pi)||$ is also strictly almost positive. Moreover, $||a_{j,j'}(\pi)||$ is a symmetric matrix and $||a_{j,j'}(\pi)|| = ||a_{j,j'}(\tilde{\pi})||$. It follows that

$$E_{\tilde{\pi}} h(T) - E_\pi\, h(T) = \sum_{j=1}^{k} \sum_{j'=1}^{k} a_{j,j'}(\pi)\{\tilde{\pi}_{1,j}\tilde{\pi}_{2,j'} - \pi_{1,j}\pi_{2,j'}\}$$

$$= \frac{1}{4} \sum_{j=1}^{k} \sum_{j'=1}^{k} a_{j,j'}(\pi)\, (\pi_{1,j}-\pi_{2,j})\, (\pi_{1,j'}-\pi_{2,j'}) > 0 \qquad (1.5)$$

because $\sum(\pi_{1,j}-\pi_{2,j}) = 0$ and $\pi_{1,j} - \pi_{2,j} \neq 0$ for some j. Hence $E_\pi\, h(T)$ can't assume its maximum at any point $\pi \neq \bar{p}$ and since it does assume its maximum, the second part of the lemma is proved.

To prove the first part we need only exhibit a function $\tilde{h}$ for which $||\tilde{h}(t + e_j + e_{j'})||$ is strictly almost positive for all $t \in R_{k,n-2}$. Then, given any h satisfying the almost positivity requirement of the lemma, $h + \delta\tilde{h}$ satisfies this requirement strictly for every $\delta > 0$ and we apply the second part of the lemma and pass to the limit as $\delta \to 0$. The simplest function $\tilde{h}$ is $\tilde{h}(t) = \exp(\frac{1}{2}|t|^2)$ where $|t|$ is the Euclidean norm of t. To show that $\tilde{h}$ works, we show more generally that $||\tilde{h}(x_j + x_{j'})||_{k\times k}$ is strictly positive for all distinct $x_1,\ldots,x_k \in \mathbb{R}^m$. Since

$$\tilde{h}(x) = (2\pi)^{-\frac{1}{2}m} \int_{\mathbb{R}^m} \exp(sx' - \tfrac{1}{2}|s|^2)ds,$$

$$\sum_{j=1}^{k} \sum_{j'=1}^{k} \tilde{h}(x_j + x_{j'})c_j c_{j'}$$

$$= (2\pi)^{-\frac{1}{2}m} \int_{\mathbb{R}^m} \left\{ \sum_{j=1}^{k} c_j \exp(sx_j') \right\}^2 \exp(-\tfrac{1}{2}|s|^2)ds \geq 0$$

with equality holding if and only if

$$\sum_{j=1}^{k} c_j \exp(sx_j') = 0 \qquad \text{a.e.} \quad \text{on } \mathbb{R}^m. \tag{1.6}$$

By the unicity of the Laplace-Stieltjes transform (1.6) implies $c_1=\ldots=c_k=0$, so that $||\tilde{h}(x_j + x_{j'})||$ is indeed strictly positive. The lemma is proved. □

We note that Hoeffding's theorem on the Poisson-binomial distribution was also proved by Gleser [6] using majorization and Schur functions. With the aid of multivariate generalizations of these concepts, Rinott [12] essentially proved lemma 1.1. The reader should note, however, that the assumption on h in lemma 1.1 is not equivalent to convexity of h, as one might infer from Rinott [12], p. 72.

Proof of Theorem 1.1. Since (i) $\Rightarrow$ (ii) is trivial it suffices to show (ii) => (iii) => (i) as well as the second part of the theorem.

(ii) => (iii). Given $x_1,\ldots,x_k \in \mathbb{R}^m$ and $\omega_1,\ldots,\omega_k$ not all 0 with $\sum \omega_j = 0$, let $\omega_j^+ = \max(\omega_j, 0) \geq 0$ and $\omega_j^- = -\min(\omega_j, 0) \geq 0$ so that $\sum \omega_j^+ = \sum \omega_j^- > 0$. Let F_i assign mass $p_{i,j}$ to x_j for $i = 1,2$, $j = 1,\ldots,k$, where

$$p_{1,j} = \omega_j^+ / \sum_{j=1}^{k} \omega_j^+, \quad p_{2,j} = \omega_j^- / \sum_{j=1}^{k} \omega_j^-.$$

Inequality (1.3) with $n = 2$ becomes in this case (c.f. (1.5))

$$0 \leq \sum_{j=1}^{k} \sum_{j'=1}^{k} g(x_j + x_{j'}) \{\tfrac{1}{4}(p_{1,j}+p_{2,j})(p_{1,j'}+p_{2,j'}) - p_{1,j}p_{2,j'}\}$$

$$= \frac{1}{4} \sum_{j=1}^{k} \sum_{j'=1}^{k} g(x_j + x_{j'})(p_{1,j}-p_{2,j})(p_{1,j'}-p_{2,j'}),$$

which is equivalent to $\sum\sum g(x_j + x_{j'})\omega_j\omega_{j'} \geq 0$, the almost positivity requirement (iii).

(iii) => (i). Given $A = \{x_1,\ldots,x_k\} \subset \mathbb{R}^m$ and $(F_1,\ldots,F_n) \in \mathcal{F}(A)$, let $p_{i,j} = F_i(\{x_j\})$ and

$$h(t) = g\left(\sum_{j=1}^{k} x_j t_j\right)$$

for $t \in R_{k,n}$. If $||g(z_j + z_{j'})||_{k\times k}$ is (strictly) almost positive for all $z_1,\ldots,z_k \in \mathbb{R}^m$, then clearly $||h(t+e_j+e_{j'})||_{k\times k}$ is (strictly) almost positive for all $t \in R_{k,n-2}$. Moreover, inequality (1.3) is equivalent to (1.4), so (iii) => (i) by the first part of lemma 1.1. In the same way the second part of the theorem follows from the second part of lemma 1.1. The theorem is proved. □

II. FURTHER ANALYSIS OF SPECIAL CASES

In this section we explore in detail for some special cases for what kind of functions $g : \mathbb{R}^m \to \mathbb{R}^1$ the assertions of theorem 1.1 hold. Define

$$\mathcal{G}_k = \{g : g \text{ measurable and (1.3) holds for all } n \text{ and } F \in \mathcal{F}_k\}$$

$$\mathcal{G}_\infty = \bigcap_k \mathcal{G}_k .$$

Of course these classes depend also on m but this is suppressed in the notation. The remarks following theorem 1.1 lead to the conclusion that, for all m,

$$\mathcal{G}_2 = \{g : g \text{ is continuous and convex on } \mathbb{R}^m\}.$$

To deal with the case $m = 1$, $k \geq 3$, we need the concept of total positivity (c.f. Karlin [9]). A function $f : \mathbb{R}^2 \to \mathbb{R}^1$ is called totally positive of order n (TP_n) if for every $r = 1,\ldots,n$, every $x_1 < x_2 <\ldots< x_r \in \mathbb{R}^1$ and every $y_1 < y_2 <\ldots< y_r \in \mathbb{R}^1$, $\det||f(x_j,y_{j'})||_{r\times r} \geq 0$; if this inequality is always strict, then f is called strictly totally positive of order n (STP_n). If f is STP_n, then in particular $\det||f(x_j,x_{j'})||_{r\times r} > 0$ for all $r = 1,\ldots,n$ and distinct $x_1,\ldots,x_n \in \mathbb{R}^1$, and this implies that the matrix $||f(x_j,x_{j'})||_{n\times n}$ is strictly positive for all distinct $x_1,\ldots,x_n \in \mathbb{R}^1$. By an approximation argument based on lemma 9.1 in Chapter 2 of Karlin [9] one finds that if f is TP_n then $||f(x_j,x_{j'})||_{n\times n}$ is positive for all $x_1,\ldots,x_n \in \mathbb{R}^1$.

We shall only be concerned with the translation case where $f(x,y) = h(x+y)$ and in this case there is also a converse. If $||h(x_j+x_{j'})||_{n\times n}$

is strictly positive for all distinct $x_1,\dots,x_n \in \mathbb{R}^1$ then theorem 8.1 of Chapter 2 of Karlin [9] ensures that $h(x+y)$ is TP_n. The approximation argument at the end of the proof of lemma 1.1 shows that the assumption that $\|h(x_j+x_{j'})\|_{n\times n}$ is positive for all $x_1,\dots,x_n \in \mathbb{R}^1$ is already sufficient to ensure that $h(x+y)$ is TP_n.

After these preliminaries we formulate

Theorem 2.1. For $m = 1$ and $k \geq 3$, G_k is the class of functions $g : \mathbb{R}^1 \to \mathbb{R}^1$ that are twice differentiable and for which $g''(x+y)$ is TP_{k-1} (or equivalently, for which $\|g''(x_j+x_{j'})\|_{(k-1)\times(k-1)}$ is positive for all $x_1,\dots,x_{k-1} \in \mathbb{R}^1$).

Proof. According to theorem 1.1, G_k is the class of measurable functions for which, for every $x_1,\dots,x_k \in \mathbb{R}^1$,

$$\sum_{j=1}^{k}\sum_{j'=1}^{k} g(x_j+x_{j'})\omega_j\omega_{j'} \geq 0 \qquad \text{if} \quad \sum_{j=1}^{k}\omega_j = 0. \tag{2.1}$$

Obviously, this is equivalent to

$$\sum_{j=1}^{k-1}\sum_{j'=1}^{k-1} \Big\{ g(x_{j+1}+x_{j'+1}) + g(x_j+x_{j'}) - g(x_j+x_{j'+1}) - g(x_{j+1}+x_{j'}) \Big\} c_j c_{j'} \geq 0 \tag{2.2}$$

for all $x_1,\dots,x_k \in \mathbb{R}^1$ and $c_1,\dots,c_{k-1} \in \mathbb{R}^1$. Define

$$\Delta_\delta^1\, g(x) = g(x+\delta) - g(x)$$

$$\Delta_\delta^r\, g(x) = \Delta_\delta^1\, \Delta_\delta^{r-1}\, g(x).$$

Taking $x_1 = \frac{1}{2}x$ and $x_{j+1} - x_j = \delta > 0$ for $j = 1,\dots,k-1$, (2.2) reduces to

$$\sum_{j=1}^{k-1} \sum_{j'=1}^{k-1} \Delta_\delta^2 \, g(x+(j+j'-2)\delta) c_j c_{j'} \geq 0 \tag{2.3}$$

for all $x \in \mathbb{R}^1$, $\delta > 0$ and $c_1,\ldots,c_{k-1} \in \mathbb{R}^1$. For $c_1 = -1$, $c_2 = 1$, $c_3 = \ldots = c_{k-1} = 0$, (2.3) implies in particular that $\Delta_\delta^4 \, g(x) \geq 0$ for all x and $\delta > 0$. But this means that g is twice differentiable with a continuous and convex second derivative g'' (cf. Popoviciu [10] or Boas and Widder [3]).

Suppose first that $g \in G_k$, so that $||\Delta_\delta^2 \, g(x+(j+j')\delta)||_{j,j'=1}^{k-1}$ is positive and g'' is continuous. For $\delta > 0$, $\varepsilon > 0$, define

$$h_{\delta,\varepsilon}(x) = \frac{1}{\delta^2} \Delta_\delta^2 \, g(x) + \varepsilon \, e^{\frac{1}{2}x^2} .$$

By the argument at the end of the proof of lemma 1.1, $||h_{\delta,\varepsilon}(x+(j+j')\delta)||_{j,j'=1}^{k-1}$ is strictly positive for every $\delta > 0$, $\varepsilon > 0$ and x. Hence, for every $\delta > 0$, $\varepsilon > 0$, x, j_0, j_0' and $r = 1,\ldots,k-1$,

$$\det||h_{\delta,\varepsilon}(x+(j+j')\delta)||_{\substack{j=j_0,\ldots,j_0+r-1 \\ j' = j_0',\ldots,j_0'+r-1}} > 0$$

and by repeated application of theorem 3.1 of Chapter 2 of Karlin [9],

$$\det ||h_{\delta,\varepsilon}(x+(j+j')\delta)||_{\substack{j=j_1,\ldots,j_r \\ j' = j_1',\ldots,j_r'}} > 0 \tag{2.4}$$

for every $\delta > 0$, $\varepsilon > 0$, x, $j_1 < j_2 < \ldots < j_r$, $j_1' < j_2' < \ldots < j_r'$ and $r = 1,\ldots,k-1$. Now choose $r \leq k-1$, $x_1 < \ldots < x_r$, $y_1 < \ldots < y_r$, and sequences $\delta_N \downarrow 0$ and $\varepsilon_N \downarrow 0$. In (2.4) substitute $x = 0$, $\delta = \delta_N$, $\varepsilon = \varepsilon_N$, $j_\nu = [x_\nu/\delta_N]$, $j_\nu' = [y_\nu/\delta_N]$ for $\nu = 1,\ldots,r$, and let $N \to \infty$.

Because g'' is continuous, the sequence $h_{\delta_N,\varepsilon_N}$ is equicontinuous and we find $\det||g''(x_j+y_{j'})||_{r\times r} \geq 0$ for every $r \leq k-1$, $x_1 <\ldots< x_r$ and $y_1 <\ldots< y_r$, which means that $g''(x+y)$ is TP_{k-1}.

Conversely, suppose that $g''(x+y)$ is TP_{k-1}. Since $k \geq 3$, we certainly have the TP_2 property $g''(x)g''(x+2\delta) \geq g^2(x+\delta)$ for all x and δ, so $\log g''$ is convex and g'' is therefore continuous. For $x_1 <\ldots< x_k \in \mathbb{R}^1$ and $c_1,\ldots,c_{k-1} \in \mathbb{R}^1$,

$$\sum_{j=1}^{k-1}\sum_{j'=1}^{k-1} \{g(x_{j+1}+x_{j'+1}) + g(x_j+x_{j'}) - g(x_j+x_{j'+1}) - g(x_{j+1}+x_{j'})\}c_jc_{j'}$$

$$= \sum_{j=1}^{k-1}\sum_{j'=1}^{k-1} c_jc_{j'} \int_{x_{j'}}^{x_{j'+1}}\int_{x_j}^{x_{j+1}} g''(\xi_j+\xi_{j'})d\xi_j d\xi_{j'}$$

$$= \prod_{\nu=1}^{k-1} (x_{\nu+1}-x_\nu)^{-1} \int_{x_{k-1}}^{x_k} \cdots \int_{x_1}^{x_2} \left[\sum_{j=1}^{k-1}\sum_{j'=1}^{k-1} g''(\xi_j+\xi_{j'})a_ja_{j'}\right] d\xi_1\ldots d\xi_{k-1}$$

where $a_j = c_j(x_{j+1}-x_j)$, $j = 1,\ldots,k-1$. The expression within square brackets is nonnegative because $g''(x+y)$ is TP_{k-1} and consequently (2.2) is satisfied so that $g \in G_k$. The proof is complete. □

Note that in the proof of theorem 2.1 we have shown that every function $g \in G_3$ has a continuous and convex second derivative g''. In a similar way one can show that for $k \geq 3$, every $g \in G_k$ is $2(k-2)$ times differentiable and its derivatives of even order $g^{(2m)}$ are continuous, nonnegative and convex for $m = 1,\ldots,(k-2)$.

For $m \geq 2$ we have no results comparable with theorem 2.1. In fact our remaining results concern G_∞ only. First of all we establish an intuitively obvious interpretation of the class G_∞.

Theorem 2.2. For every m, G_∞ is the class of measurable functions $g : \mathbb{R}^m \to \mathbb{R}^1$ for which (1.3) holds for all n and all $F = (F_1,\ldots,F_n)$ such that $E_{\overline{F}}\, g(S_n)$ exists.

Proof. If g satisfies (1.3) for all n and F such that $E_{\overline{F}}\, g(S_n)$ exists, then clearly $g \in G_\infty$. Conversely, suppose that $g \in G_\infty$ and consider, first of all, an $F = (F_1,\ldots,F_n)$ for which all F_i, $i = 1,\ldots,n$, have compact support, i.e. $F_i(K) = 1$ for a compact set $K \subset \mathbb{R}^m$ and $i = 1,\ldots,n$. For $k = 1,2,\ldots$, let $F^{(k)} = (F_1^{(k)},\ldots,F_n^{(k)})$, where $F_i^{(k)}(A_k) = 1$ for $i = 1,\ldots,n$, $A_k \subset K$ and card $(A_k) = k$, and such that $F_i^{(k)}$ converges weakly to F_i for $i = 1,\ldots,n$ as $k \to \infty$. Clearly such a sequence $\{F^{(k)}\}$ exists and has the further property that $F_\cdot^{(k)} = n^{-1} \sum F_i^{(k)}$ also has $F_\cdot^{(k)}(A_k) = 1$ and converges weakly to $F_\cdot$ which has $F_\cdot(K) = 1$. Since $F^{(k)} \in F_k$, we have $E_{F(k)}\, g(S) \le E_{\overline{F}(k)}\, g(S)$ for all $k = 1,2,\ldots$, and since $g \in G_\infty \subset G_2$, g is continuous and the weak convergence of $F_1^{(k)}$ and $F_\cdot^{(k)}$ yield $E_F\, g(S) \le E_{\overline{F}} g(S)$.

Now suppose only that $E_{\overline{F}}\, g(S)$ exists but that $F = (F_1,\ldots,F_n)$ is otherwise arbitrary. Note that this implies that for any $i_1,\ldots,i_n \in \{1,\ldots,n\}$ and $G = (F_{i_1},\ldots,F_{i_n})$, $E_G\, g(S)$ exists and in particular that $E_F\, g(S)$ exists. Let K be a compact set in $\mathbb{R}^m$ and define $F^* = (F_1^*,\ldots,F_n^*)$ by $F_i^*(B) = F_i(K \cap B)/F_i(K)$ for $i = 1,\ldots,n$, and $F_\cdot^* = n^{-1} \sum F_i^*$, $\overline{F}^* = (F_\cdot^*,\ldots,F_\cdot^*)$ accordingly. In view of the first part of the proof, $E_{F^*}\, g(S) \le E_{\overline{F}^*}\, g(S)$ which may be written as

$$\left[\prod_{i=1}^{n} F_i(K)\right]^{-1} \int \cdots \int_{K^n} g\left(\sum_{i=1}^{n} x_i\right) \prod_{i=1}^{n} d\, F_i(x_i)$$

$$\le n^{-n} \sum_{i_1=1}^{n} \cdots \sum_{i_n=1}^{n} \left[\prod_{\nu=1}^{n} F_{i_\nu}(K)\right]^{-1} \int \cdots \int_{K^n} g\left[\sum_{i=1}^{n} x_i\right] \prod_{\nu=1}^{n} d\, F_{i_\nu}(x_\nu).$$

Now let $K \uparrow \mathbb{R}^m$ so that $F_i(K) \to 1$ for $i = 1,\ldots,n$, and apply the monotone convergence theorem separately to the positive and negative parts of the integrands in each of the above integrals. Since $E_{\underline{F}}\, g(S)$ and $E_F\, g(S)$ exist, this yields

$$E_F\, g^+(S) - E_F\, g^-(S) \leq E_{\underline{F}}\, g^+(S) - E_{\underline{F}}\, g^-(S)$$

and we have proved (1.3) and the theorem. □

Let us now turn to a description of the class G_∞. For $m = 1$, theorem 2.1 implies that G_∞ consists of those functions $g : \mathbb{R}^1 \to \mathbb{R}^1$ for which $g''(x+y)$ is totally positive of every order. According to Karlin [9], p.77, this is equivalent to g'' being a Laplace transform of a nonnegative sigma-finite measure ν on $\mathbb{R}^1$. Hence for $m = 1$, G_∞ is the class of functions g for which

$$g''(x) = \int_{\mathbb{R}^1} e^{ux}\, \nu(du). \tag{2.5}$$

For $m > 1$, we don't have an analogue of theorem 2.1 and we have to approach the problem in a different manner. We shall prove

Theorem 2.3. For every m, G_∞ is the class of functions $g : \mathbb{R}^m \to \mathbb{R}^1$ which are of the form

$$g(x) = c + \mu x' + \frac{1}{2} x \sum x' + \int_{0<|u|\leq 1} (e^{ux'} - 1 - ux')\nu(du) \tag{2.6}$$

$$+ \int_{|u|>1} e^{ux'}\nu(du),$$

where $c \in \mathbb{R}^1$, $\mu \in \mathbb{R}^m$, Σ is a positive $m \times m$ matrix and ν a nonnegative sigma-finite measure on $\mathbb{R}^m - \{0\}$ such that

$$\int_{0<|u|\leq 1} |u|^2 \nu(du) < \infty, \tag{2.7}$$

$$\int_{|u|>1} e^{\lambda|u|} \nu(du) < \infty \quad \text{for every } \lambda \geq 0. \tag{2.8}$$

Proof. If (2.6), (2.7) and (2.8) hold, then clearly $|g(x)| < \infty$ for all $x \in \mathbb{R}^m$. For every k, every $x_1,\ldots,x_k \in \mathbb{R}^m$ and every $\omega_1,\ldots,\omega_k$ with $\sum \omega_j = 0$,

$$\sum_{j=1}^{k} \sum_{j'=1}^{k} g(x_j+x_{j'})\omega_j\omega_{j'} = \left(\sum_{j=1}^{k} \omega_j x_j\right) \Sigma \left(\sum_{j=1}^{k} \omega_j x_j\right)'$$

$$+ \int_{R^m-\{0\}} \left(\sum_{j=1}^{k} \omega_j e^{ux_j'}\right)^2 \nu(du) \geq 0$$

because Σ is positive. It follows that $g \in G_\infty$.

By a theorem of Löwner and Schoenberg (cf. Donoghue [4], p. 135) the $k \times k$ matrix $||g(x_j+x_{j'})||$ is almost positive for every k and every $x_1,\ldots,x_k \in \mathbb{R}^m$ if and only if $||\exp\{tg(x_j+x_{j'})\}||_{k\times k}$ is positive for every k, every $x_1,\ldots,x_k \in \mathbb{R}^m$ and every $t \geq 0$. But this is equivalent to $\exp\{tg\}$ being a Laplace transform of a nonnegative measure on $\mathbb{R}^m$ for every $t \geq 0$: for $m = 1$ this is a classical result (cf. Widder [13], p. 273, and Boas and Widder [3]); for $m > 1$ there is a highly incomplete proof in Akhiezer [1], pp. 229-231; a full proof was

given recently by Reeds [11]. Hence G_∞ consists of those functions g for which $\exp\{g(x) - g(0)\}$ is the Laplace transform of an infinitely divisible probability measure on $\mathbb{R}^m$.

Let Z have an infinitely divisible probability distribution on $\mathbb{R}^m$. By the Lévy-Khinchine formula (cf. Gikhman and Skorokhod [5], theorems 1,2, pp. 271-273) this is equivalent to

$$Z = \mu + Z_1 + Z_2 + Z_3,$$

where Z_1, Z_2, Z_3 are independent, Z_1 possesses an m-dimensional $N(0,\sum)$ distribution and Z_2 and Z_3 have characteristic functions

$$E\, e^{ixZ_2'} = \exp\left\{ \int_{0<|u|\leq 1} (e^{iux'} - 1 - iux')\nu(du)\right\}, \tag{2.9}$$

$$E\, e^{ixZ_3'} = \exp\left\{ \int_{|u|>1} (e^{iux'} - 1)\nu(du)\right\} . \tag{2.10}$$

Here μ and $\sum$ are as in the formulation of the theorem and ν is a nonnegative measure satisfying (2.7) and $\nu(\{u:|u|>\varepsilon\}) < \infty$ for every $\varepsilon > 0$. In this setting, $g \in G_\infty$ if and only if $E \exp\{xZ_r'\} < \infty$ for $r = 1,2$ and all $x \in \mathbb{R}^m$, and

$$\exp\{g(x) - g(0)\} = \exp\{\mu x' + \tfrac{1}{2} x \textstyle\sum x'\}\, E\, e^{xZ_2'}\, E\, e^{xZ_3'} . \tag{2.11}$$

Suppose that $g \in G_\infty$ so that (2.11), (2.9), (2.10) and (2.7) are satisfied, $\nu(|u| > \varepsilon) < \infty$ for every $\varepsilon > 0$ and $E \exp\{xZ_r'\} < \infty$. Since $\int_{0<|u|\leq 1} (\exp\{ux'\} - 1 - ux')\nu(du) < \infty$ for all x by (2.7) we conclude by analytic continuation that

$$E\, e^{xZ_2'} = \exp\left\{ \int_{0<|u|\leq 1} (e^{ux'} - 1 - ux')\nu(du)\right\} . \tag{2.12}$$

Finally, assume without loss of generality that $\nu(|u| > 1) > 0$, define for Borel sets $A \subset \mathbb{R}^m$

$$Q(A) = \frac{\nu(A \cap \{|u| > 1\})}{\nu(|u| > 1)},$$

let $Y_1, Y_2, \ldots$ be independent and identically distributed according to Q and let N be independent of $Y_1, Y_2, \ldots$ and have a Poisson distribution with parameter $\nu(|u| > 1)$. Now Z_3 is distributed as $\sum_{i=1}^{N} Y_i$ and we conclude that

$$E\, e^{xY_1'} \le E\, e^{xZ_3'} / P(N=1) < \infty.$$

Thus $\int_{|u|>1} \exp\{ux'\}\nu(du) < \infty$ for all $x \in \mathbb{R}^m$. Therefore condition (2.8) is satisfied and

$$E\, e^{xZ_3'} = \exp\left\{ \int_{|u|>1} (e^{ux'} - 1)\nu(du) \right\}. \tag{2.13}$$

Together (2.11), (2.12) and (2.13) yield (2.6) and the theorem follows.□

The simplest examples of functions satisfying the conditions of theorem 2.3 are functions g which are themselves Laplace transforms of finite (not necessarily infinitely divisible) measures on $\mathbb{R}^m$ and of course $g(x) = |x|^2$ and (trivially) the linear functions.

One may also consider the problems discussed in this paper for functions defined only on an additively closed subset of $\mathbb{R}^m$ rather than on all of $\mathbb{R}^m$. For $m = 1$ for instance, let us consider continuous functions g defined on $(0,\infty)$ and ask when (1.3) holds for all $F = (F_1, \ldots, F_n)$ with $F_i(0,\infty) = 1$, $i = 1, \ldots, n$, and all n. By essentially the same analysis as above we find that g must be the logarithm of the Laplace transform of a (not necessarily finite) non-negative infinitely divisible measure on $\mathbb{R}^1$; of course this Laplace transform need only be finite on $(0,\infty)$. By theorem 4.2 of Horn [8] this holds if and only if g'' is of the form (2.5) on $(0,\infty)$, or, equivalently, if

$$g(x) = g(x_0) + b(x-x_0) + \int_{\mathbb{R}^1} \left[e^{ux} - e^{ux_0} - u(x-x_0)e^{ux_0}\right]u^{-2}\nu(du) \tag{2.14}$$

for any $x_0 > 0$, $b \in \mathbb{R}^1$ and ν a nonnegative measure for which the integrals in (2.5) and (2.14) are finite for all $0 < x < \infty$. We can for instance apply this result to conclude that $g(x) = x^{-1}$, $x > 0$, satisfies (1.3) when the X_i are restricted to be positive.

It is a pleasure to acknowledge our indebtedness to J. Fabius and R.A. Olshen for helping to improve our understanding of some of the problems discussed in this paper and to H.P. Wynn for bringing Donoghue [4] to our attention.

REFERENCES

[1] Akhiezer, N.I. (1965). The Classical Moment Problem and Some Related Questions in Analysis. Oliver and Boyd, Edinburgh.

[2] Berg, C., Christensen, J.P.R. and Ressel, P. (1976). Positive definite functions on Abelian semigroups. Math. Ann. 223, 253-274.

[3] Boas, R.P.,Jr. and Widder, D.V. (1940). Functions with positive differences. Duke Math. J. 7, 496-503.

[4] Donoghue, W.F.,Jr. (1974). Monotone Matrix Functions and Analytic Continuation. Springer, Berlin.

[5] Gikhman, I.I. and Skorokhod, A.V. (1969). Introduction to the Theory of Random Processes. Saunders, Philadelphia.

[6] Gleser, L.J. (1975). On the distribution of the number of successes in independent trials. Ann. Probability 3, 182-188.

[7] Hoeffding, W. (1956). On the distribution of the number of successes in independent trials. Ann. Math. Statist. 27, 713-721.

[8] Horn, R.A. (1967). On infinitely divisible matrices, kernels and functions. Z. Wahrscheinlichkeitstheorie verw. Geb. 8, 219-230.

[9] Karlin, S. (1968). Total Positivity. Vol. I. Stanford University Press, Stanford.

[10] Popoviciu, T. (1945). Les Fonctions Convexes. Hermann, Paris

[11] Reeds, J. (1979). Widder theorem characterization of Laplace transforms of positive measures on $\mathbb{R}^k$. Technical Report, Department of Statistics, University of California, Berkeley.

[12] Rinott, Y. (1973). Multivariate majorization and rearrangement inequalities with some applications to probability and statistics. Israel J. Math. 15, 60-77.

[13] Widder, D.V. (1946). The Laplace Transform. Princeton University Press, Princeton.

SEQUENTIAL MINIMUM PROBABILITY RATIO TESTS

W. J. Hall[1]

Department of Statistics
University of Rochester
Rochester, New York

A family of sequential tests, including SPRT's as a special case, is described which permit control on a weighted average of the error probabilities; when sufficient symmetry conditions obtain, control on each error probability is achieved. The method of test construction is essentially implicit in a paper of Wassily Hoeffding's (1960), in which he developed a lower bound on the ASN of sequential tests at an intermediate hypothesis. A subset of them have been independently introduced by Lorden (1976) - his 2-SPRT's. Examples include tests for a normal mean, binomial tests, t-tests, Stein-analog tests, robust tests, and exponential family tests. A multiple-decision identification problem is also considered. In the normal mean case, a member of this class is almost identical to a test proposed by Anderson (1960).

I. INTRODUCTION

In this paper we describe a family of sequential test procedures, some examples of which have been known for almost twenty years (e.g., Hall, 1960), but which have not been explored in full generality in the literature. This occasion is especially appropriate since this family of tests has its origins in a paper of Wassily Hoeffding's (1960), where they are essentially implicit. A full description has never appeared in the literature, however, although examples are now well-known to many students of sequential analysis. Recently, Lorden (1976) did give a

[1]Supported by the U.S. Army Research Office through the University of Rochester.

ISBN: 0-12-166650-6

brief description of an important subset of them and proved an asymptotic optimality property. We shall give a much fuller description of the tests, describe a number of examples, indicate a multiple-decision extension, and finally sketch their origin by reviewing parts of Hoeffding's 1960 paper; Lorden's result is also summarized.

In order to be true to the title of this conference, a brief review of asymptotic versions of these tests will be given; but a fuller account will be published separately, along with tables of their asymptotic characteristics (Lambert & Hall, 1979).

Let us consider the simple example of sequentially testing whether $N(\theta,1)$-observations have mean $\pm\ \delta$. If we stop the first time the smaller of the likelihood ratios of one of these hypotheses to a $N(0,1)$ hypothesis is at most 2α, with terminal decision by maximum likelihood, we can show that each of the two error probabilities is at most α; the test is carried out by plotting the cumulative sum of the observations against the sample size, and using certain symmetric converging straight-line barriers. We refer to the test as an MPRT. Anderson (1960) also proposed linear stopping barriers for this problem, and he computationally determined them so as to minimize the ASN when $\theta = 0$. Lorden (1976) has since proved that the MPRT (his 2 -SPRT) is asymptotically optimal, in that the difference between its ASN (at $\theta = 0$) and the smallest possible ASN for a test with the same error probabilities α, tends to zero with α.

We describe a larger family of such tests, generated by allowing other intermediate hypotheses (not necessarily iid), by relaxing the symmetry of this example, and by introducing weights on the likelihood ratios. This can be done in a way to maintain control on a weighted average of the error probabilities. When the role of the intermediate hypothesis is played by a mixture of the two hypotheses under test, the

MPRT reduces to an SPRT. Some curious new bounds on averages of the error probabilities of SPRT's are thereby derived. Various other examples are treated.

Perhaps the most promising example described herein is the robust sequential test for the location parameter of a 'near normal' distribution, based on techniques of Huber (1965, 1968). The observations (appropriately centered) are first censored: x is replaced by $x^* = \text{med}(-c,x,c)$ for some positive c. Then the cumulative sum is plotted sequentially until it crosses one of two converging straight-line barriers, as above. Simple formulas for the parameters of this stopping region are given so that each error probability is at most α whenever the sampled population is within a specified neighborhood of $N(\pm\, \delta, \sigma^2)$ (δ, σ specified), or stochastically more extreme. The practical drawback of having to specify the variance σ^2 of the 'core hypotheses' remains, but asymptotically it can presumably be replaced by a (robust) estimate, much as in the Stein-analog of the non-robust normal mean problem (Hall, 1962).

Another potentially interesting example is the m-decision identification problem, in which one achieves error control under a 'least favorable configuration' in which all populations are identical except for a single best population. Here, we have a sequential procedure alternative to that given by Bechhofer, Kiefer & Sobel (1968), though still 'open' in the sense that there is no upper bound on the sample size. However, arguing by analogy with the two-decision case, it seems likely that this alternative procedure tends to have a smaller maximum ASN than the SPRT-like procedure of Bechhofer, Kiefer & Sobel.

II. MINIMUM PROBABILITY RATIO TESTS

Consider a sequence of experiments with the following notation and assumptions:

X_n:	data from stage n $(n = 1,2,\ldots)$
f_{in} $(i = 0,1,2)$:	distinct alternative joint densities of $X_{(n)} \equiv (X_1,\ldots,X_n)$ (w.r.t. a common dominating measure μ_n), consistently defined for $n = 1,2,\ldots$
H_i $(i = 0,1,2)$:	hypothesis that $\{f_{in}\}$ is correct
d_i $(i = 1,2)$:	terminal decision in favor of H_i
λ_i $(i = 1,2)$:	positive weights (or prior probabilities), adding to unity
$\ell_n = \min_{i=1,2}(\lambda_i f_{in}/f_{0n})$ $= \ell_n(X_1,\ldots,X_n)$	($\equiv 0$ if either numerator vanishes.)

We are interested in sequentially testing the simple hypotheses H_1 vs. H_2; H_0 is subsidiary, and should be thought of as an 'intermediate' hypothesis. We refer to ℓ_n as the 'minimum (weighted) probability ratio.'

Define a sequential test procedure (N,D), referred to as a <u>minimum probability ratio test</u> (MPRT), as follows:

<u>stopping rule N</u>: N is the smallest positive integer n for which $\ell_n \leq \alpha$ (< 1).

<u>terminal decision rule D</u>: maximum (weighted) likelihood - choose d_1 (d_2, resp.) if $\lambda_1 f_{1n} > (<, \text{resp.})\ \lambda_2 f_{2n}$, with an arbitrary choice if equality holds.

That is, we stop whenever $\lambda_1 f_{1n} \le \alpha f_{0n}$ or $\lambda_2 f_{2n} \le \alpha f_{0n}$, and decide d_2 or d_1 accordingly. The class of MPRT's of H_1 vs. H_2 is generated by choices of the subsidiary hypothesis H_0, the weights $\underline{\lambda} = (\lambda_1, \lambda_2)$, and the bound α; we write $\text{MPRT}(\alpha, \underline{\lambda}, H_0)$.

In the special case when H_0, as well as H_1 and H_2, are iid hypotheses, Lorden (1976) called such tests <u>2-SPRT's</u>; they may be viewed as a <u>union</u> of two SPRT's. Specifically, let N_i be the stopping time for an SPRT of H_i vs. H_0 based on f_{in}/f_{0n} with stopping barriers $A_i > B_i > 0$ $(i= 1,2)$. If we let $B_i = \alpha/\lambda_i$ and A_i be sufficiently large to be irrelevant, then the MPRT has $N = \min(N_1, N_2)$: continue until at least one of the SPRT's stops. By contrast, the Armitage (1947) and Sobel & Wald (1949) 'two-sided' test procedure (for testing H_0 vs. H_1 H_2) may be described as the <u>intersection</u> of two SPRT's, with $N = \max(N_1, N_2)$: continue until both SPRT's stop. (Thus, in the normal mean example, the MPRT boundaries are the 'inner boundaries' of the Armitage and Sobel-Wald procedure; see below.)

We proceed to derive the basic inequalities for the $\text{MPRT}(\alpha, \underline{\lambda}, H_0)$:

$$\lambda_1 P_1(d_2) \le \alpha P_0(d_2), \quad \lambda_2 P_2(d_1) \le \alpha P_0(d_1), \quad \text{and} \tag{1}$$

$$\lambda_1 P_1(d_2) + \lambda_2 P_2(d_1) \le \alpha \tag{2}$$

where P_i represents probability under H_i. Thus, the <u>weighted average of the error probabilities is at most α</u> (pre-specified); in particular, with $\lambda_1 = \lambda_2$, we have a pre-specified bound (2α) on the sum of the error proabilities.

To verify (1) and (2), let S_{in} be the subset of the range of $X_{(n)}$ in which $N = n$ and d_i is chosen. In S_{2n}, $\lambda_1 f_{1n} \le \lambda_2 f_{2n}$ so that $\lambda_1 f_{1n} = \ell_n f_{0n} \le \alpha f_{0n}$. Hence

$$\lambda_1 P_1(d_2) = \sum_n \int_{S_{2n}} \lambda_1 f_{1n} \le \sum_n \int_{S_{2n}} \alpha f_{0n} = \alpha P_0(d_2)$$

and similarly with subscripts '1' and '2' interchanged, proving (1); adding the RHS's in (1) yields $\alpha P_0(N < \infty) \le \alpha$, whence (2). Equality would hold in (1) and (2) except for 'excess' (ℓ_N may be less than α) and lack of 'termination' under H_0 ($P_0(N < \infty)$ may be less than unity).

III. RELATIONSHIP WITH SPRT's

In this section we confine attention to H_0's of the form

$$f_{0n} = \mu_1 f_{1n} + \mu_2 f_{2n} \quad (\mu_i > 0,\ \mu_1 + \mu_2 = 1,\ \underline{\mu} = (\mu_1,\mu_2)); \tag{3}$$

and we write $\text{MPRT}(\alpha,\underline{\lambda},\underline{\mu})$. We now show that this family of MPRT's includes all SPRT's; we write SPRT(B,A) for the SPRT of H_1 vs. H_2 which stops in favor of $H_1(H_2)$ when $\ell_n \le B$ $(\ge A)$, for given $0 < B < A$.

<u>PROPOSITION 1</u>: Given $0 < B < A$ and writing $r(t) = t/(1+t)$, for every λ in the interval $I \equiv (r(B), r(A))$, there exists $(\alpha,\underline{\lambda},\underline{\mu})$ for which

$$\text{MPRT}(\alpha,\underline{\lambda},\underline{\mu}) \equiv \text{SPRT}(B,A); \tag{4}$$

specifically, $\lambda_1 = \lambda$, $\lambda_2 = 1-\lambda$, $\mu_1 = a(\lambda)/[a(\lambda)+b(\lambda)]$ where $a(\lambda) = B(1+A)[A(1+A)^{-1}-\lambda]$ and $b(\lambda) = (1+B)[\lambda-B(1+B)^{-1}]$, $\mu_2 = 1-\mu_1$, and $\alpha = \lambda/[A + (1-A)\mu_1]$. Moreover, (2) holds for every such λ.

Hence, for each SPRT, there are many convex combinations of error probabilities with known upper bound (with equality except for excess). The two extremal cases, with $\lambda \downarrow r(B)$ or $\uparrow r(A)$, yield $BP_1(d_2) + P_2(d_1) \le B$ and $AP_1(d_2) + P_2(d_1) \le 1$, which, when treated as equations, yield Wald's approximations. The 'intermediate' case of $\lambda = 1/2$ (requiring $B < 1 < A$), yields $P_1(d_2) + P_2(d_1) \le (AB - 2B + 1)/(A-B)$ which simplifies to $\alpha + \beta$ when $A = (1-\beta)/\alpha$ and $B = \beta/(1-\alpha)$ - another of Wald's inequalities.

Proof of Proposition 1: We have $\ell_n \le B$ iff $\mu_2\ell_n/(\mu_1+\mu_2\ell_n) \le \mu_2B/(\mu_1+\mu_2B)$. The LHS of the latter inequality is μ_2f_{2n}/f_{0n} (see (3)) and hence equating the RHS to α identifies one of the SPRT stopping rules with one of those of the MPRT. Similarly, considering $\ell_n \ge A$ leads to equating $\lambda_1/(\mu_1+\lambda_2A)$ to α. (The terminal decision rules are identical since λ_1 in I iff $B < \lambda_1/\lambda_2 < A$.) Solving these two equations for μ_1 and α as functions of $\lambda_1 = \lambda$, A and B yields the formulas in Proposition 1. It is easy to check that $\mu_1 \in (0,1)$ iff $\lambda \in I$, and that μ_1 is decreasing in λ; therefore α is between λ and $\lambda/A < 1/(1+A)$, and hence in $(0,1)$.

In fact, the class of $MPRT(\alpha,\underline{\lambda},\underline{\mu})$'s is essentially equivalent to the class of SPRT's:

PROPOSITION 2: Given the $MPRT(\alpha,\underline{\lambda},\underline{\mu})$ with $\alpha < \lambda_1\lambda_1/(\lambda_1\mu_2 + \lambda_2\mu_1)$ $(<(\lambda_1/\mu_1)\wedge(\lambda_2/\mu_2))$, there exists $0 < B < A$ for which (4) holds; specifically, $A = (\lambda_1-\alpha\mu_1)/(\alpha\mu_2)$ and $B = \alpha\mu_1/(\lambda_2 - \alpha\mu_2)$.

The proof is elementary; the condition on α assures $0 < B < A$.

Writing $MPRT(\alpha,\underline{\lambda})$ for $MPRT(\alpha,\underline{\lambda},\underline{\lambda})$, we have the resulting simpler equivalences: Given $0 < B < A$, choose $\lambda_1 = (AB)^{\frac{1}{2}}/[1+(AB)^{\frac{1}{2}}]$, $\lambda_2 = 1-\lambda_1$, $\alpha = (AB)^{1/2}/[A+(AB)^{1/2}]$, and then

$$MPRT(\alpha,\underline{\lambda}) \equiv SPRT(B,A). \qquad (5)$$

Given $(\alpha,\underline{\lambda})$ with $\alpha < 1/2$, choose $A = \rho\gamma$, $B = \rho/\gamma$ where $\rho = \lambda_1/\lambda_2$ and $\gamma = (1-\alpha)/\alpha$, and then (5) holds.

IV. THE PARAMETRIC CASE

Suppose $(X_1,\ldots,X_n)$ have joint density $g_{\theta n}$ for some real θ, and H_i: $\theta = \theta_i (\theta_1 < \theta_0 < \theta_2)$; thus H_0 is an intermediate hypothesis. (More generally, we could let $f_{0n} = \int g_{\theta n} dG(\theta)$ for some df G; we have already considered the possiblity that G is discrete on $\{\theta_1, \theta_2\}$ with masses μ_1 and μ_2, leading to an SPRT.)

If $\{g_{\theta n}\}$ is a MLR family in $T_n = t_n(X_1,\ldots,X_n)$, then any such MPRT is of the form: stop as soon as either $T_n \leq b_n$ or $\geq a_n$ and decide in favor of d_1 (d_2) if $T_n < c_n$ $(> c_n)$, where c_n is between a_n and b_n. Since $g_{\theta_2 n}/g_{\theta_1 n}$ is monotone in T_n, the MPRT is a GSPRT (see Weiss, 1953): stop in favor of H_2 if $g_{\theta_2 n}/g_{\theta_1 n} \geq A_n$, or H_1 if $\leq B_n$ $(B_n \leq A_n)$, and continue sampling otherwise. As a consequence, this MPRT has a monotone OC function (Ghosh, 1960), and may be considered as a test of $\theta \leq \theta_1$ vs. $\theta \geq \theta_2$.

V. THE SYMMETRIC CASE

Let T be a 1-1 (measurable) transformation of the probability space taking H_1 into H_2 and conversely (i.e., if $X_{(n)}$ has density f_{1n}, then $(TX)_{(n)}$ has f_{2n} and $(TTX)_{(n)}$ has f_{1n}), and T leaves H_0 invariant. Let $\lambda_1 = \lambda_2 = 1/2$. (For convenience, assume $f_{1n} = f_{2n}$ with probability 0, or else randomize between d_1 and d_2 when $f_{1N} = f_{2N}$.)

The MPRT is now symmetric in the idices '1' and '2', and invariant in '0', so that $P_0(d_1) = P_0(d_2) \leq 1/2$ $(= 1/2$ if $P_0(N < \infty) = 1)$. The basic inequalities (1) become

$$P_1(d_2) \leq \alpha \quad \text{and} \quad P_2(d_1) \leq \alpha, \tag{6}$$

with equalities except for 'excess' and 'termination' under H_0.

The symmetric MPRT, denoted simply $MPRT(\alpha,H_0)$, simplifies to:

$$\text{stop whenever } \min(f_{1n}/f_{0n}, f_{2n}/f_{0n}) \le 2\alpha \tag{7}$$

and choose between H_1 and H_2 by maximum likelihood - i.e., stop and choose d_2 if $f_{1n} \leq 2\alpha f_{0n}$ or d_1 if $f_{2n} \le 2\alpha f_{0n}$. If H_0 specifies $f_{0n} = (f_{1n} + f_{2n})/2$, then $MPRT(\alpha, H_0) = SPRT(B,A)$ with $A = (1-\alpha)/\alpha = 1/B$, a symmetric SPRT. We now consider several symmetric examples, with other choices of H_0.

Example 1 (normal mean): The X_j's are iid $N(\theta,1)$, $\theta_1 < \theta_2$. Choose θ_0 midway between. W.l.o.g., let $\theta_1 = -\delta$, $\theta_0 = 0$, $\theta_2 = \delta(> 0)$. We readily find the MPRT (5) to be:

$$\text{stop whenever } |S_n| \equiv |\textstyle\sum_{j=1}^n X_j| \ge -\delta^{-1}\log(2\alpha) - \frac{1}{2}\delta n \equiv a^\circ - b^\circ n \tag{8}$$

and choose d_2 or d_1 according as S_N is positive or negative. Hence, the stopping barriers in the (n,S_n)-plane are converging straight lines $\pm(a^\circ - b^\circ n)$ (Hall, 1960). Note that $N \le [n_0+]$ (smallest integer $\ge n_0$) where $n_0 = -2\delta^{-2}\log(2\alpha)$. Each error probability is at most α. (Lorden suggests replacing α in a° by $\alpha/(.9992 - .5729\delta + .1392\delta^2)$ to adjust for excess; the resulting error probabilities then equal α within 1%, for α between .1% and 10% and δ between .1 and 1.) This test is one of Anderson's (1960) tests. Asymptotic formulas for the (monotone) OC function and ASN function appear in Anderson (1960); see Lambert & Hall (1979).

The symmetry is here obvious: the transformation T changes signs of all the X_j's. Alternative symmetric tests may be obtained by letting G be a symmetric distribution for θ, and taking $f_{0n} = \int \pi_{i=1}^n g_\theta dG(\theta)$ where g_θ is the $N(\theta,1)$ density. Specifically, suppose G assigns probability 1/2 each to $\pm\eta$. Then N is the smallest n for which

$$\delta|S_n| + \log\cosh(\eta S_n) \geq -\log(2\alpha) - \frac{1}{2} n(\delta^2 - \eta^2); \tag{9}$$

termination occurs with certainty under H_0 so long as $|\eta| \leq \delta$. With $\eta = 0$, we get the MPRT (8); with $\eta = \delta$, the SPRT; intermediate values give 'intermediate' sequential tests, with $N \leq [-2(\delta^2-\eta^2)^{-1} \log(2\alpha)+2$. Or letting G be $N(0,1/m)$, we find N to be the smallest n for which

$$|S_n| \geq -r\delta + [-2r \log(2\alpha) + rm\delta^2 + r \log(r/m)]^{1/2} \text{ where } r = m + n. \tag{10}$$

All of these tests have error probabilities at most α; little else is known except for the test in (8).

Since 'inequality' is due solely to 'excess', the continuous-time Wiener process version has error probabilities exactly equal to α. The Wiener process versions of (9) and (10) lead to unusual moving barrier absorption problems for Wiener processes with drift $\pm\,\delta$, with absorption probabilities equal to α and $1 - \alpha$. See section 8.

Tests for the mean of a finite population, both SPRT and MPRT, appear in Hall (1978).

Example 2 (Bernoulli): The X_j's are iid Bernoulli θ, with $\theta_1 = (1-\delta)/2$, $\theta_2 = (1+\delta)/2$, $\theta_0 = 1/2 (0 < \delta < 1)$. If the Bernoulli variables X_j are replaced by $Y_j = 2X_j - 1$, the symmetry transformation is again a sign change. One readily finds that the MPRT (7) may be carried out by plotting $S_n = \sum_{j=1}^n Y_j$ vs. n, as in (8), with the stopping boundary intercepts $a^\circ = -2 \log(2\alpha)/\log[(1+\delta)/(1-\delta)]$ and $-a^\circ$, and slopes $-b^\circ = -\log(1-\delta^2)/\log[(1+\delta)/(1-\delta)]$ and b° - again, converging straight line boundaries (Hall, 1960). See Lorden (1977) for some examples. As in Example 1, other symmetric boundaries are possible, intermediate between these and the horizontal SPRT boundaries. All have error probabilities at most α.

Example 3 (t-test): Let Y_j's be iid $N(\mu,\sigma^2)$, $\theta = \mu/\sigma$, H_1: $\theta = -\delta$, H_0: $\theta = 0$, H_2: $\theta = +\delta(\delta>0)$. By an invariance reduction, let X_j be the t-statistic based on $(Y_1,\ldots,Y_j)$ $(j > 1)$, and consider tests based on the sequence of X_j's. Now the joint likelihood ratio of successive t-statistics (for any pair of θ-values) is monotone in the last t-statistic (by invariance and sufficiency considerations - see Hall, Wijsman & Ghosh (1965) - and by the MLR property of the non-central t-distribution).

Hence, MLR as well as symmetry conditions prevail and the MPRT may be described by a diagram in the (n,X_n)-plane, with an upper and a lower (symmetric) stopping barrier. (The barriers are curved lines.)

The test may be made explicit by calculating the likelihood ratio of non-central t $(=x_n)$, with non-centrality $\pm\delta$, to central t, and stopping whenever the smaller of these is at most 2α; see HWG (1965) for formulas.

Note that this is not the usual one-sided t-test problem; here we are testing whether $\mu/\sigma = \pm\delta$ instead of $\mu = 0$ vs. $\mu/\sigma = \Delta(>0)$. An asymptotically valid MPRT for testing these latter hypotheses appears in Hall (1973b).

Example 4 (normal mean, variance unknown): This problem, like Example 3 except H_i is $\mu = \pm\delta$ or 0 rather than $\mu/\sigma = \pm\delta$ or 0, was treated in Hall (1962) and is not repeated here.

Example 5 (robust test): The X_j's are iid F. Let F_1 be $N(-\delta,1)$ and F_2 be $N(\delta,1)$, and let $H_1 = \{F|F(x) \geq \bar{\varepsilon}\, F_1(x) - \varepsilon'$ for all $x\}$, $H_2 = \{F|\bar{F}(x) > \bar{\varepsilon}\, F_2(x) - \varepsilon'$ for all $x\}$ where $\bar{\varepsilon} = 1-\varepsilon$, $\bar{F} = 1-F$, and ε,ε' are small positive constants; H_1 is the hypothesis that F is 'near F_1' or stochastically larger, H_2 that F is 'near F_2' or stochastically smaller. With $\varepsilon' = 0$, 'near' can be interpreted in the contamination or gross error senses. This hypothesis testing problem was considered by Huber (1968); see also Huber (1965).

Huber found least favorable members of H_i, say F_i^* with density f_i^*, and constructed tests of H_1 vs. H_2 by using likelihood ratio tests, nonsequential or sequential, of f_1^* vs. f_2^*. With error probabilities controlled at f_i^*, the error probabilities are nowhere larger throughout H_i. The tests turn out to be based on cumulative sums of censored observations X_j^*.

First, consider the family of densities $g(x|\theta,\eta)$ which are continuous, proportional to $N(\theta,1)$ in the middle (for $|x| \le c$), and proportional to an equal mixture of $N(\pm\,\eta,1)$ in each tail; explicitly, write

$$g(x|\theta,\eta) = \phi(x^*-\theta)[\psi(x,\eta)/\psi(x^*,\eta)]/k(\theta,\eta)$$

where ϕ is the $N(0,1)$ density, $x^* = \text{med}(-c,x,c)$ ('x censored to $[-c,c]$'), $\psi(x,\eta) = \phi(x-\eta) + \phi(x+\eta)$, $\eta \ge 0$, $k(\cdot,\cdot)$ an integration constant (even in θ); the positive parameter c is omitted from the notation. At $c = \infty$, $g(x|\theta,\eta) = \phi(x-\theta)$.

Huber's least favorable f_i^*'s can conveniently (for our purposes) be expressed as: $f_1^* = g(\cdot|-\delta,\delta)$, $f_2^* = g(\cdot|\delta,\delta)$ with c related to δ, ε and ε' through the equation

$$\overline{\Phi}(c-\delta) - \overline{\Phi}(c+\delta)e^{2\delta c} = \varepsilon[\overline{\Phi}(c-\delta) + \Phi(c+\delta)e^{2\delta c}] + \varepsilon'(1+e^{2\delta c}).$$

(For small δ, and correspondingly small ε and ε', we find $h(c)\delta \doteq \varepsilon + 2\varepsilon'$ where $h(c) = 2[\phi(c) - c\overline{\Phi}(c)]$; e.g., $c = 0.5$, 1.0, $1,5$, 2.0 yields $h(c) = .40$, $.17$, $.06$, $.02$.) Since $g(\cdot|\theta,\eta)$ is an exponential family in θ, inference about θ for given η – e.g., testing f_1^* vs. f_2^* – is based on $\sum X_i^* = S_n^*$; in fact, the log likelihood ratio of f_2^* to f_1^* is $2\delta S_n^*$, on which an SPRT of H_1 vs. H_2 may be based (Huber (1965, 1968)).

Now let H_0^*: F has density $f_0^* = g(\cdot|0,\delta)$. Then H_0^*, H_1^*, H_2^* satisfy the symmetry conditions under sign change, so we can construct a symmetric MPRT.

We shall need the integration constants $k(\delta,\delta)$ and $k(0,\delta)$; the first turns out to be unity and we find for the second:

$$k(0,\delta) = \frac{2\,\phi(c)}{\psi(c,\delta)}[2 - \Phi(c-\delta) - \Phi(c+\delta)] + 2\Phi(c) - 1$$

(which correctly equals unity at $c = \infty$); for later use we also note that $\log k(0.\delta) = 1/2\ \zeta_c\delta^2 + o(\delta^4, (c\delta)^4)$ where $\zeta_c = 2[c\phi(c) - (c^2-1)\overline{\Phi}(c)]$, in (0,1) and decreasing.

Applying (7) with f_{in}^*, we find analogously to (8):

$$\begin{aligned} \text{stop whenever } |S_n^*| &\geq -\delta^{-1}\log(2\alpha) - 1/2\ n\delta + n\delta^{-1}\log k(0,\delta) \\ &\doteq -\delta^{-1}\log(2\alpha) - 1/2\ n\delta(1-\zeta_c) \text{ for small } \delta. \end{aligned} \tag{11}$$

Comparing with (8), (11) is based on the sum S_n^* of censored observations (at $\pm$ c) and the converging straight line barriers are less steep. Each of these facts tends to increase the sample size, but of course the error probabilities at F_1 and F_2 have also been reduced; they are at most α at the least favorable F_1^* and no larger for any F (including F_i) in H_i. For example, for testing $\delta = \pm 1/6$ with $\varepsilon' = 0$ and $c = 1.5$, we allow for a contamination level of $\varepsilon \doteq 1\%$, with error probabilities maintained at α. More details will be given elsewhere (hall, 1973a).

Example 6 (correlation): With iid bivariate normal observations, consider testing $\rho = \pm\, \delta$ $(0 < \delta < 1)$, using $\rho = 0$ as the symmetry point. A test may be based on the invariantly sufficient statistic r, the sample correlation coefficient. Details are not given here, since it is of dubious interest; by contrast, the usual one-sided test is $\rho = 0$ vs. $\rho = \Delta(>0)$.

Example 7 (identification): Suppose the X_j's are independent and (for convenience only) identically distributed. Each X_j is of the form $X_j = (U_j, V_j)$ where U and V each have the same dimension. According to H_1 (U,V) has joint density f (specified), while according to H_2 (V,U) has joint density f; H_0 specifies that (U,V) has a specific exchangeable joint density g. The transformation T exchanges the coordinates of (U,V). In other words, the U's and V's have known distributions, but the 'labels have been lost.' (Actually, we've generalized slightly, permitting U and V to be dependent.) The testing problem is to determine the 'labels.' Any symmetric (exchangeable) g can be used to construct an MPRT with known (and equal) error probability bounds; however, one would want to choose a g with some care, to assure termination.

A special case is as follows: Suppose U and V independent and each with a density from the exponential family

$$f(x|\theta) = a(x)e^{x\theta - h(\theta)} \ (d\nu) \tag{12}$$

for θ in a real interval; H_1 implies (U,V) have parameters (θ_1, θ_2) and H_2 implies (θ_2, θ_1), while H_0 implies (θ_0, θ_0). And let $\theta_1 = \theta_0 - \delta$, $\theta_2 = \theta_0 + \delta(\delta > 0)$, and $Y_j = V_j - U_j$. We find that the MPRT is given by the diagram of Example 1, plotting S_n vs. n, with upper decision line $a_n = a^o - b^o n$, lower decision line $b_n = -a_n$, with a^o as before and $b^o = \delta^{-1}[h(\theta_0+\delta) + h(\theta_0-\delta) - 2h(\theta_0)]$.

Some extensions are possible, using invariance to allow for some unknown parameters. We return to identification problems in Section 7.

VI. ASYMMETRIC CASE - EXPONENTIAL FAMILY

We now leave the symmetric case, and confine attention to the univariate exponential family.

Suppose the X_j's are iid with marginal density given by (12), with $\theta_1 < \theta_0 < \theta_2$. The MPRT is found to be:

$$\text{stop if either } S_n \equiv \sum X_j \geq a_1^o - b_1^o n \equiv a_n \text{ or } \leq -a_2^o + b_2^o n \equiv b_n \quad (13)$$

and choose $d_2(d_1)$ if $S_n > (<) \pi_1 a_n + \pi_2 b_n \equiv c_n$ where $a_i^o = -\delta_i^{-1} \log(\alpha/\lambda_i)$, $b_i^o = \delta_i^{-1}(h_i - h_0)$, $h_i = h(\theta_i)$, $\delta_i = |\theta_i - \theta_0|$, $\pi_i = \delta_i/(\theta_2-\theta_1) = \delta_i/(\delta_1+\delta_2)$. We shall require θ_0 to be such that $a_1^o > -a_2^o$ and $b_1^o > -b_2^o$ so that the test has converging straight line boundaries in the (n,S_n)-plane. The sample size is bounded above by $[n_o +]$, where $n_o = (\pi_2 \log \lambda_1 + \pi_1 \log \lambda_2 - \log \alpha)/(\pi_2 h_1 + \pi_1 h_2 - h_0)$. Note that $b_n < c_n < a_n$ for $n < n_0$. (The above includes many of the examples already treated when most of the symbols above do not depend on i so that symmetry obtains.)

Without symmetry, we can only bound a weighted average of the error probabilities (2); there is no apparent way to choose θ_0 so as to determine upper bounds on the individual error probabilities.

Asymptotically (see below), S_n will behave like a Wiener process with drift $h'(\theta)$. It is readily verified that, if θ_0 is fixed and $\theta_i = \theta_0 \pm \delta$ $(\delta \downarrow 0)$, then (with $\lambda_1 = \lambda_2 = 1/2$) $a_1^o = a_2^o$ and $b_1^o = b_2^o + O(\delta^2)$ $(b_i = O(\delta))$, and therefore $P_0(d_i)$ is approximately 1/2 by asymptotic symmetry. Then each error probability is approximately bounded above by α. But that θ_0 should be exactly midway between θ_1 and θ_2 non-asymptotically seems unlikely. Expansion of $P_0(d_i)$, for small $\theta_2-\theta_1$, may reveal an appropriate choice of θ_0. In the case of testing hypotheses about the Poisson mean, Fukushima (1961) compared two choices of θ_0 by direct computation.

VII. MULTIPLE-DECISION EXTENSION - IDENTIFICATION PROBLEM

Suppose X_j has m components $(Y_{1j},\ldots,Y_{mj})$ and denote $Y_{i(n)} = (Y_{i1},\ldots,Y_{in})$. According to $H_i (i = 1,\ldots,m)$, $X_{(n)} \equiv (Y_{i(n)},\ldots,Y_{m(n)})$ is assumed to have the joint density

$$f_{in} = h_n(y_{i(n)}) \prod_{j\neq i} g_n(y_{j(n)})$$

while according to H_0 $X_{(n)}$ has an exchangeable joint density $f_{0n}(y_{1(n)},\ldots,y_{m(n)})$ - invariant under permutation of the $y_{j(n)}$'s.

Consider the stopping rule N: stop whenever $\lambda_n < m\alpha$ where

$$\ell_n = \min_i \sum_{j\neq 1} f_{jn}/f_{0n}$$

and the terminal decision rule D: Decide d_i - that is, choose H_i - $(i = 1,\ldots,m)$ according to maximum likelihood, or, equivalently, 'minimum unlikelihood' (Hall, 1958): decide d_i according to the smallest $\sum_{j\neq 1} f_{jn}$ (with arbitrary but symmetric treatment of ties).

Hence, in the stopping region for H_i, say S_{in}, $\sum_{j\neq 1} f_{jn} \leq m\alpha f_{0n}$. By symmetry, $P_i(d_j)$ is independent of i and j $(i\neq j)$ and $P_0(d_i)$ is independent of i. If $P_1(N < \infty) = 1$, we therefore have

$$P_i(d_i) = 1 - \sum_{j \neq i} P_i(d_j) = 1 - \sum_{j \neq i} P_j(d_i)$$

$$= 1 - \sum_n \int_{S_{in}} \sum_{j \neq i} f_{jn} \geq 1 - \sum \int_{S_{in}} m\alpha f_{0n}$$

$$= 1 - m\alpha P_0(d_i) \geq 1 - \alpha.$$

(Equality holds in the last step iff $P_0(N < \infty) = 1$). Hence, each probability of a correct decision is controlled, and likewise each error probability $P_i(d_j)$ is at most $(m-1)^{-1} \alpha$ $(i \neq j)$.

This identification problem may be described as that of picking the best population when one of the m populations is uniquely best and all others are identical - frequently a 'least favorable configuration.' This and a variety of other identification problems have been considered by Bechhofer, Kiefer & Sobel (1968). We now proceed to show, when all populations are members of an exponential family ('Koopman-Darmois family'), that Bechhofer, Kiefer & Sobel's sequential procedure is included among those described above - it is the m-decision analog of the SPRT, obtained by choosing H_0 to be a kind of 'average' of the H_i's. We then develop an alternative procedure to theirs, an m-decision analog of what in the 2-decision case was a converging straight line MPRT, with H_0 'intermediate' to the other H_i's. We limit attention, however, to this very special identification problem because of its symmetry; sequential procedures can also be developed for asymmetric versions, but as in the 2-decision case, control on individual error probabilities, or correct decision probabilities, is not possible.

We assume all observations have densities of the form (12). According to H_i, Y_{ij} has parameter $\theta_1 = \theta_0 + \delta$ and Y_{kj} $(k \neq i)$ has parameter $\theta_0 (\delta > 0)$. Then

$$f_{in} = c \exp[\theta_1 S_{in} + \theta_0 \sum_{j \neq i} S_{jn} - nh_1 - n(m-1)h_0]$$

where $h_i = h(\theta_i)$ and c is a symmetric function of the y_{ij}'s. Setting $f_{0n} = m^{-1}\sum_i f_{in}$ (as we did in the 2-decision case to obtain an SPRT, with $\mu_i = \lambda_i = 1/2$), we find

$$\ell_n = m \sum_{j=1}^{m-1} \exp(\delta S_{(j)n}) / \sum_{j=1}^{m} \exp(\delta S_{(j)n})$$

where $S_{(1)n} \le \ldots \le S_{(m)n}$ are the ordered S_{in}'s. Then $\ell_n \le m\alpha$, the criterion for stopping according to N, occurs iff

$$\sum_{j=1}^{m-1} \exp[-\delta(S_{(m)n} - S_{(j)n})] \le \alpha/(1-\alpha), \tag{14}$$

in agreement with Bechhofer, Kiefer & Sobel's equation (4.3.8), page 93. The corresponding terminal decision rule D is: choose the population with the largest S_{in}.

If instead we now choose H_0 as specifying all observations are iid with $\theta = \theta^* = \theta_0 + m^{-1}\delta$ (so that the average θ is the same for every H_i, $i = 0,1,\ldots,m$), then $f_{in} = c \exp(\theta^* m\overline{S}_n - nmh^*)$, where $\overline{S}_n = m^{-1}\sum_i S_{in}$ and $h^* = h(\theta^*)$, and

$$\ell_n = \sum_{j=1}^{m=1} \exp[\delta S_{(j)n} + (\theta_0 - \theta^*)m\overline{S}_n - n\Delta]$$

where $\Delta \equiv h_1 + (m-1)h_0 - mh^*$. We find the stopping rule N becomes: stop the first time

$$\sum_{j=1}^{m-1} \exp[-\delta(\overline{S}_n - S_{(j)n})] \le [\exp(n\Delta)]m\alpha. \tag{15}$$

In the $N(\theta,1)$ case, $\Delta = 1/2\ \delta^2$ (and equals $1/2\ \sigma^2\delta^2$ in the asymptotic case of Section 8). Again, terminal decision is by maximum likelihood.

The procedure (15), for $m > 2$, is not 'closed' - the stopping time is not bounded. It may be shown to terminate with certainty, however, but other properties are not known.

The above two procedures may be extended to the t-test case (Example 3), using invariance and sufficiency and carefully preserving the symmetry, and the 'Stein analog' case (Example 4), but the properties have not been explored.

VIII. ASYMPTOTIC VERSION

Continuing with the exponential family example of Section 6, suppose we now write $\theta = \theta_0 + \gamma A^{-1}\delta$, $\theta_1 = \theta_0 - \delta$ (i.e., $\gamma = -A$) and $\theta_2 = \theta_0 + \delta(\gamma = A)$ where $A = [-2\log(2\alpha)]^{1/2}$ and $\lambda_1 = \lambda_2$. We examine this case (which is asymptotically symmetric) as $\delta \downarrow 0$.

Recall that $E_\theta X = h'(\theta)$ and $\mathrm{var}_\theta X = h''(\theta)$. By a location shift, we may and do suppose $h'(\theta_0) = 0$ and denote $h''(\theta_0) = \sigma^2$. Then $h(\theta) = h(\theta_0) + 1/2\ \sigma^2\gamma^2A^{-2}\delta^2 + o(\delta^2)$, $h'(\theta) = \sigma^2\gamma A^{-1}\delta + o(\delta)$, $h''(\theta) = \sigma^2 + o(\delta)$, $h(\theta_1) + h(\theta_2) - 2h(\theta_0) = \sigma^2\delta^2 + o(\delta^2)$ and $n_o = A^2/[\sigma^2\delta^2 + o(\delta^2)] = n_o(\delta)$.

Let $W_\gamma(\delta,t) = n_o(\delta)^{-1/2} \sum_{j=1}^{[n_o t]} X_j/\sigma = n_o(\delta)^{-1/2}\sum(X_j - h'(\theta_0+\varepsilon))/\sigma + n_o^{-1/2}[n_o t]h'(\theta_0+\varepsilon)/\sigma$ where $\varepsilon = \delta\gamma/A$. The first term converges weakly (as $\delta \downarrow 0$) to a standard Wiener process $W(t)$ on $[0,1]$, and the second term tends to γt (uniformly in t). Hence, $W_\gamma(\delta,t)$ converges weakly to a Wiener process $W_\gamma(t)$ with drift γ, on $[0,1]$.

The boundaries of the MPRT (13), when re-scaled for $W_\gamma(\delta,t)$, are asymptotically $\pm \frac{1}{2} A(1-t)$. Hence, the asymptotic version is the following <u>"canonical" Wiener process MPRT</u>:

We begin observing $W_\gamma(t)$, a Wiener process on [0,1] with drift γ and unit variance, and stop when $W_\gamma(t)$ hits $\pm \frac{1}{2} A(1-t)$ $(A>0)$. The upper (lower) boundaries are associated with terminal decisions in favor of H_2: $\gamma = A$ (H_1: $\gamma = -A$). Each error probability is $\alpha = 1/2 \exp(\delta - 1/2\, A^2)$.

The OC and ASN functions (each functions of γ, with the single parameter A) may be obtained from formulas in Anderson (1960); further details and tables appear in Lambert & Hall (1979).

IX. HOEFFDING'S LOWER BOUND AND THE MPRT; LORDEN'S PROBABILITY

Hoeffding (1960) introduced a lower bound on the ASN of any test of two simple hypotheses H_1 and H_2 when another hypothesis H_0 is true. It presumes (as we do throughout this section) that the observations are iid under all three hypotheses. He went on to show that the lower bound is 'almost achieved' by one of Anderson's tests, closely related to what we have called the MPRT for a normal mean (Example 1 above) when the error probabilities are .05 and .01. Moreover, following through Hoeffding's argument, the symmetric MPRT (for iid H_0) is essentially implicitly defined, as a contender for near-optimality. For error probabilities near zero or near one-half, however, other tests were shown to have ASN's under H_0 near his bound.

Lorden (1976) obtained stronger results. His 2-SPRT(A,B) (which is our MPRT with his (A,B) our $(\alpha/\lambda_1, \alpha/\lambda_2)$) was shown to be asymptotically optimal in minimizing the ASN at H_0 - in a surprising strong sense. Lorden proved the following theorem (recast in our notation): Let (α_1, α_2) be the true error probabilities of the MPRT $(\alpha, \underline{\lambda}, f_0)$ of H_1 vs. H_2 with H_0 specifying iid f_0 observations, and let $N = N(\alpha, \underline{\lambda}, f_0)$ be its stopping time. Let $n = n(\alpha, \underline{\lambda}, f_0) = \{\inf E_0 N |$ all tests with error probabilities at most $(\alpha_1, \alpha_2)\}$.

THEOREM (Lorden): Under assumptions (A) and (D) below, $E_0N - n \to 0$ as $\alpha \to 0$. (Actually, he only requires $(\alpha/\lambda_1) \wedge (\alpha/\lambda_2) \to 0$.) This is a strong result in that each term on the LHS is tending to infinity while the difference is tending to zero. It is surprising in that Hoeffding showed (in the symmetric normal mean case) that the sample size of a suitable nonsequential test $\sim n_0$ as $\alpha \to 0$ where n_0 is his lower bound on n, and hence that a nonsequential test is near optimal for sufficiently small α. He also showed that a SPRT was near optimal for α near 1/2. Further comments appear below.

Hoeffding's basic result is simpler than Lorden's. We sketch his argument for what it reveals, and to demonstrate the implicit definition of the MPRT $(\alpha,\underline{\lambda},f_0)$ $(\lambda_1 = \lambda_2 = 1/2)$. We list the following assumptions:

(A) f_0, f_1, f_2 are distinct and mutually absolutely continuous.

(B) $\rho_i \equiv E_0\log[f_i(X)/f_0(X)] > -\infty$ $(i = 1,2)$ (and < 0 by Jensen).

(C) $\tau^2 \equiv \mathrm{var}_0\log[f_1(X)/f_2(X)] < \infty$ (and $\tau > 0$).

(D) $E_0\log^2[f_i(X)/f_0(X)] < \infty$ $(i = 1,2)$ (implying (B) and (C)).

Let (N,D) be a sequential test of H_1 vs. H_2 with error probabilities $\alpha_1 = P_1(d_2)$ and $\alpha_2 = P_2(d_1)$.

THEOREM (Hoeffding): Under (A) - (C),

$$E_0N \geq [(\lambda^2 + K)^{1/2} - \lambda]^2 \tag{16}$$

where $K = \rho^{-1}\log\min(\alpha_1 + \alpha_2, 1)$, $\rho = \rho_1 \wedge \rho_2$, and $\lambda = -\tau/(4\rho)$. (Note that $\rho < 0$, $\lambda > 0$, $K \geq 0$. The inequality (16) is trivial unless $\alpha_1 + \alpha_2 < 1$ and $E_0N < \infty$, so we assume these hereafter.)

The bound (16) depends on the error probabilities only through their sum, and any upper bound (< 1) on $\alpha_1 + \alpha_2$ may be substituted if the error probabilities are not exactly known. Recall that such an upper bound is available for the MPRT (with $\lambda_1 = \lambda_2 = 1/2$), and also for the SPRT(B,A), using Wald's stopping boundaries $A = (1-\alpha_2')/\alpha_1'$ and $B = \alpha_2'/(1-\alpha_1')$ (the bound being $\alpha_1' + \alpha_2'$).

<u>Proof</u>: Letting S_{in} (i=1,2) denote the stopping set for d_i, we have

$$\begin{aligned}\alpha_1 + \alpha_2 &= \sum_n \int_{S_{2n}} f_{1n} + \sum_n \int_{S_{1n}} f_{2n} \\ &\geq \sum_n \int_{S_{2n}} \min_{i=1,2} f_{in} + \sum_n \int_{S_{1n}} \min_{i=1,2} f_{in} \\ &= \sum_n \int_{\{N=n\}} [\min(f_{in}/f_{0n})] f_{0n} \qquad \text{(using (A))} \\ &= E_0 \min(f_{in}/f_{0n}) \qquad \text{(since } P_0(N<\infty) = 1).\end{aligned} \tag{17}$$

So far, then,

$$\begin{aligned}\log(\alpha_1 + \alpha_2) &\geq \log E_0 \min(f_{iN}/f_{0N}) \\ &\geq E_0 \log \min(f_{iN}/f_{0N}) \text{ (Jensen)}\end{aligned} \tag{18}$$

$$= E_0 \min \textstyle\sum_{j=1}^N \log[f_i(X_j)/f_0(X_j)]$$

$$= E_0 \min(Z_{iN} + N\rho_i) \quad \text{where} \quad Z_{in} = \textstyle\sum_{j=1}^n \{\log[f_i(X_j)/f_0(X_j)] - \rho_i\}$$

$$\geq E_0(\min Z_{in} + N\rho) \tag{19}$$

$$= E_0 \min Z_{in} + \rho E_0 N = -1/2\, E_0 Z_N + \rho E_0 N$$

(where $Z_n = |Z_{1n} - Z_{2n}|$ since $\min(a,b) = 1/2(a+b) - 1/2|a-b|$ and $E_0 Z_{iN} = 0$ by Wald's equation and (B))

$$\geq -1/2[E_0 Z_N^2]^{1/2} + \rho E_0 N. \tag{20}$$

But $E_0 Z_N^2 = E_0(Z_{1N} - Z_{2N})^2 = E_0[\sum_{j=1}^{N}(\log[f_1(X_j)/f_2(X_j)] - \rho_1 + \rho_2]^2 = E_0 N \cdot \tau^2$ by Wald's second equation and (C). We thus have $\log(\alpha_1 + \alpha_2) \geq -\frac{\tau}{2}(E_0 N)^{1/2} + \rho E_0 N$, i.e., $K \leq 2\lambda(E_0 N)^{1/2} + E_0 N$, which is equivalent to (16).

Inequalities were introduced in (17) - (20). These inequalities are, respectively, equalities iff

(17) D is maximum likelihood,

(18) $\min(f_{iN}/f_{0N})$ is constant a.s. (H_0),

(19) $\rho_1 = \rho_2$,

(20) Z_N is constant a.s. (H_0).

The first holds for any MPRT (with $\lambda_1 = \lambda_2$) or SPRT (with $A > 1 > B$), or a suitable FSST when there is sufficient symmetry. The third holds under modest symmetry conditions. The second holds, except for excess, precisely for the MPRT (since we stop when the ratio is at most 2α), and not for any SPRT or FSST; it is in this sense that the MPRT may be considered implicit in Hoeffding's development. The fourth holds, except for excess, precisely for a symmetric ($\rho_1 = \rho_2$, $B = 1/A$) SPRT, and not for the MPRT or FSST. Hence, with symmetry conditions, four sources of inequality contribute to (16); three are approximate equalities for the MPRT, three for the SPRT. Since the SPRT cannot be expected to have a small $E_0 N$, the MPRT might well be conjectured to be near optimal in minimizing $E_0 N$.

In the symmetric normal mean example, (16) reduces to

$$\delta^2 E_0 N > [(1+A^2)^{1/2} - 1]^2 \quad \text{with} \quad A^2 = -2\log(2\alpha) \; (\alpha_1 = \alpha_2 = \alpha < \tfrac{1}{2}). \tag{21}$$

Hoeffding goes on to show that, for $\alpha \uparrow 1/2$ $(A \downarrow 0)$, the leading terms of an expansion of this lower bound agree with corresponding terms in

$\delta^2 E_0 N$ for the SPRT. He also shows that, for $\alpha \downarrow 0$ ($A \uparrow \infty$), both the bound and $\delta^2 E_0 N$ for a FSST $\sim -2 \log \alpha$. Hoeffding's results, together with calculations of Anderson's for a test approximating the MPRT and $\alpha = .01$ or $.05$, support the conjecture that the MPRT may be near optimal in an intermediate range of α-values. Computations by Lambert & Hall (1979) for the Wiener process case show that the ratio of the RHS to the LHS in (21) always exceeds 96%, demonstrating the near-optimality of the MPRT uniformly in α (and δ). Lorden's computations for the normal case show even more: relative efficiencies exceed 99% for α between .0025 and .40 and δ between .1 and 1.

Some comparable calculations for a Poisson testing problem were reported by Fukushima (1961); he also compared the standard deviation of N for the SPRT and MPRT. Additional comparisons for the Wiener process case, including standard deviations and percentage points of N, appear in Lambert & Hall (1979).

REFERENCES

Anderson, T.W. (1960). A modification of the sequential probability ratio test to reduce the sample size. Ann. Math. Statist. 31, 165-197.

Armitage, P. (1947). Some sequential tests of Student's hypothesis. J. R. Statist. Soc. Suppl. 9, 250-263.

Bechhofer, Robert E., Kiefer, Jack and Sobel, Milton (1968). Sequential Identification and Ranking Procedures, University of Chicago Press.

Fukushima, Kozo (1961). A comparison of sequential tests for a Poisson parameter. M. Sc. thesis, Department of Statistics, University of North Carolina, Chapel Hill.

Ghosh, J.K. (1960). On some properties of sequential t-tests. Calcutta Statist. Assoc. Bull. 9, 139-144.

Hall, W.J. (1958). Most economical multiple-decision rules. Ann. Math. Statist. 29, 1079-1094.

Hall, W.J. (1960). A new class of sequential decision rules for symmetric problems. Ann. Math. Statist. 31, 524-525 (abstract #3). Also Invited Address at the Eastern Regional Meeting of the Inst. Math. Statist., Ithaca, NY, 1961.

Hall, W.J. (1962). Some sequential analogs of Stein's two-stage test. Biometrika 49, 367-378.

Hall, W.J. (1973a). Asymptotic efficiency of Huber's robust tests, non-sequential and sequential Proceedings of the 39th Session of the International Statistical Institute, Vol. 1, Vienna, Austria. pp. 431-432. (Fuller paper in mss form.)

Hall, W.J. (1973b). Two asymptotically efficient sequential t-tests. Proceedings of the Prague Symposium on Asymptotic Statistics, Vol. II, ed. Jaroslav Hajek, Prague, Czechoslovakia, pp. 89-102.

Hall, W.J. (1974). Asymptotic sequential analysis. Special Invited Address at the Annual Meeting of the Inst. Math. Statist., Edmonton, Alberta, Canada (to appear).

Hall, W.J. (1978). A sequential test for the mean of a finite population Invited Address at the Annual Meeting of the Inst. Math. Statist., San Diego, Calif. (to appear).

Hall, W.J., Wijsman, R.A., & Ghosh, J.K. The relationship between sufficiency and invariance with applications in sequential analysis. Ann. Math. Statist. 36, 575-614.

Hoeffding, Wassily (1960). Lower bounds for the expected sample size and the average risk of a sequential procedure. Ann. Math.Statist. 31, 352-368.

Huber, P.J. (1965). A robust version of the probability ratio test. Ann. Math. Statist. 36, 1753-1758.

Huber, P.J. (1968). Robust confidence limits. Z. Wahr. theorie und Verw. Gebiete 10, 269-278.

Lambert, D. and Hall, W.J. (1979). Tables for Wiener process sequential tests. In preparation.

Lorden, Gary (1976). 2-SPRT's and the modified Kiefer-Weiss problem of minimizing an expected sample size. Ann. Statist. 4, 281-291.

Lorden, Gary (1977). Nearly-optimal sequential tests for finitely many parameter values. Ann. Statist. 5, 1-21.

Sobel, M. and Wald, A. (1949). A sequential decision procedure for choosing one of three hypotheses concerning the unknown mean of a normal distribution. Ann. Math. Statist. 20, 502-522.

Weiss, L. (1953). Testing one simple hypothesis against another. Ann. Math. Statist. 24, 273-281.

ACKNOWLEDGMENT

I have been privileged, in ways unique among the participants of this Symposium. The writings of Wassily Hoeffding have inspired us all. But Wassily Hoeffding was my teacher; he was my dissertation supervisor; and he was my faculty colleague for ten years. In each of these roles, he has been an inspiration to me, and I am grateful.